Synthesis and Application of Luminescent Materials

Synthesis and Application of Luminescent Materials

Guest Editor
Binbin Chen

Basel • Beijing • Wuhan • Barcelona • Belgrade • Novi Sad • Cluj • Manchester

Guest Editor
Binbin Chen
School of Chemistry &
Molecular Engineering
East China University of
Science and Technology
Shanghai
China

Editorial Office
MDPI AG
Grosspeteranlage 5
4052 Basel, Switzerland

This is a reprint of the Special Issue, published open access by the journal *Inorganics* (ISSN 2304-6740), freely accessible at: www.mdpi.com/journal/inorganics/special_issues/H3P8E6X10I.

For citation purposes, cite each article independently as indicated on the article page online and using the guide below:

Lastname, A.A.; Lastname, B.B. Article Title. *Journal Name* **Year**, *Volume Number*, Page Range.

ISBN 978-3-7258-2846-3 (Hbk)
ISBN 978-3-7258-2845-6 (PDF)
https://doi.org/10.3390/books978-3-7258-2845-6

© 2024 by the authors. Articles in this book are Open Access and distributed under the Creative Commons Attribution (CC BY) license. The book as a whole is distributed by MDPI under the terms and conditions of the Creative Commons Attribution-NonCommercial-NoDerivs (CC BY-NC-ND) license (https://creativecommons.org/licenses/by-nc-nd/4.0/).

Contents

Binbin Chen
Luminescent Materials with Advanced Properties and Applications
Reprinted from: *Inorganics* 2024, *12*, 306, https://doi.org/10.3390/inorganics12120306 1

Ya-Ting Gao, Shuai Chang, Bin-Bin Chen and Da-Wei Li
Dual-Exciting Central Carbon Nanoclusters for the Dual-Channel Detection of Hemin
Reprinted from: *Inorganics* 2023, *11*, 226, https://doi.org/10.3390/inorganics11060226 4

Yujia Cheng, Yan Huang and Guang Yu
N-S-co-Doped Carbon Dot Blue Fluorescence Preparation and Baicalein Detection
Reprinted from: *Inorganics* 2024, *12*, 154, https://doi.org/10.3390/inorganics12060154 14

Karthiga Anpalagan, Hong Yin, Ivan Cole, Tian Zhang and Daniel T. H. Lai
Quantum Yield Enhancement of Carbon Quantum Dots Using Chemical-Free Precursors for Sensing Cr (VI) Ions
Reprinted from: *Inorganics* 2024, *12*, 96, https://doi.org/10.3390/inorganics12040096 29

Xiang Li, Yuxin Zang, Jindi Zhang, Lili Zhang, Jing Zhang and Mengyang Huang et al.
Orange Peel Biochar–CdS Composites for Photocatalytic Hydrogen Production
Reprinted from: *Inorganics* 2024, *12*, 156, https://doi.org/10.3390/inorganics12060156 42

Rieko Ishibashi, Ruka Koike, Yoriko Suda, Tatsuhiro Kojima, Toshiyuki Sumi and Toshiyuki Misawa et al.
Lanthanide-Containing Polyoxometalate Crystallized with Bolaamphiphile Surfactants as Inorganic–Organic Hybrid Phosphors
Reprinted from: *Inorganics* 2024, *12*, 146, https://doi.org/10.3390/inorganics12060146 57

Dimitrios Glykos, Athanassios C. Tsipis, John C. Plakatouras and Gerasimos Malandrinos
Synthesis, Spectroscopic Characterization, and Photophysical Studies of Heteroleptic Silver Complexes Bearing 2,9-Bis(styryl)-1,10-phenanthroline Ligands and Bis[(2-diphenylphosphino)phenyl] Ether
Reprinted from: *Inorganics* 2024, *12*, 131, https://doi.org/10.3390/inorganics12050131 71

Miguel Martínez, David Dalmau, Olga Crespo, Pilar García-Orduña, Fernando Lahoz and Antonio Martín et al.
Different Patterns of Pd-Promoted C-H Bond Activation in (Z)-4-Hetarylidene-5(4H)-oxazolones and Consequences in Photophysical Properties
Reprinted from: *Inorganics* 2024, *12*, 271, https://doi.org/10.3390/inorganics12100271 87

Ju Hyun Oh, Yookyoung Lee, Jihee Kim, Woo Tae Hong, Hyun Kyoung Yang and Mijeong Kang et al.
Effect of Synthesis Conditions on the Photoluminescent Properties of Si-Substituted $CaYAlO_4$:Eu: Sources of Experimental Errors in Solid-State Synthesis
Reprinted from: *Inorganics* 2024, *12*, 150, https://doi.org/10.3390/inorganics12060150 103

Lei Zhang, Mingze Xia, Yuan Zhang, Li Song, Xiwei Guo and Yong Zhang et al.
The Effect of Organic Spacer Cations with Different Chain Lengths on Quasi-Two-Dimensional Perovskite Properties
Reprinted from: *Inorganics* 2023, *12*, 12, https://doi.org/10.3390/inorganics12010012 112

Konstantin N. Gorbachenya, Elena A. Volkova, Victor V. Maltsev, Victor E. Kisel, Diana D. Mitina and Elizaveta V. Koporulina et al.
Growth, Spectroscopic Characterization and Continuous-Wave Laser Operation of Er,Yb:GdMgB$_5$O$_{10}$ Crystal
Reprinted from: *Inorganics* **2024**, *12*, 240, https://doi.org/10.3390/inorganics12090240 . . . . . . **123**

Editorial
Luminescent Materials with Advanced Properties and Applications

Binbin Chen

Key Laboratory for Advanced Materials, Shanghai Key Laboratory of Functional Materials Chemistry, Frontiers Science Center for Materiobiology & Dynamic Chemistry, School of Chemistry & Molecular Engineering, East China University of Science and Technology, Shanghai 200237, China; binbinchen@ecust.edu.cn

Luminescent materials have attracted significant attention due to their exceptional properties, which have been widely used in various fields such as sensing [1–3], bioimaging [4–6], catalysis [7–9], and optoelectronics [10–12]. Synthesis strategies for luminescent materials include the hydro/solvothermal method and microwave-assisted synthesis, as well as low-energy sustainable preparation strategies [13–15]. The structural design and optical control of luminescent materials are current research hotspots, promoting the sustained development of luminescent materials in a wide range of fields.

This Special Issue covers various luminescent materials that are currently a focus of research, such as carbon dots, perovskites, metal complexes, lanthanide phosphors, and luminescent hybrid materials, exploring their photophysical properties and achieving promising applications in chemical sensing and photocatalysis.

Advanced synthesis technology can stimulate the design and development of high-performance luminescent materials. Our group [16] developed a kind of Schiff base strategy to achieve room-temperature synthesis of carbon nanoclusters. The proposed carbon nanoclusters display a unique property of dual-exciting central emission. These nanoclusters can act as dual-channel fluorescence nanoprobes for the reliable determination of hemin based on an inner filter effect. Furthermore, Maltsev's group [17] prepared a transparent single crystal (Er^{3+},Yb^{3+}:$GdMgB_5O_{10}$) with a size of up to 24 × 15 × 12 mm through a high-temperature solution growth technique using dipped seeds. Laser operation in continuous-wave mode can be achieved, and the maximal output power is up to 0.15 W, with a slope efficiency of 11%.

Carbon dots, as a new type of luminescent carbon nanomaterial, exhibit fascinating optical properties. Yin's group [18] utilized amine-rich soybean flour (nitrogen source) and lemon juice (acidic medium) to improve the luminescent efficiency of carbon quantum dots. The enhanced quantum yield is attributed to the fact that the obtained carbon quantum dots undergo a thorough hydrothermal reaction and have zwitterionic surfaces. Meanwhile, the carbon quantum dots can be used for the specific detection of Cr(VI) ions, with a detection limit of 8 ppm. Like nitrogen-doping, sulfur-doping can also regulate the electronic structure of carbon dots. In light of this, Yu's group [19] developed blue-emitting S, N-co-doped carbon dots by using hydrothermal methods, achieving the rapid and sensitive detection of baicalein with a detection limit of 33 nM by means of static quenching and an inner filter effect.

Luminescent hybrid materials show superior photophysical properties by integrating the properties of different structural units. Ito's group [20] synthesized inorganic–organic hybrid phosphors by means of the hybridization of Eu^{3+}-containing polyoxometalate anions with bolaamphiphile surfactants. The proposed phosphors display a characteristic red emission originating from Eu^{3+} ions, with a lifetime in the order of milliseconds. Meanwhile, the emission intensity of the phosphors is laser-power-dependent, and the emission intensity increases linearly as the excitation laser power rises. Moreover, hybrid structures can also achieve efficient photocatalytic activity because the heterojunctions with an intimate interface can promote photogenerated charge transfer. Wang's group [21] constructed biochar-supported cadmium sulfide composites for photocatalytic hydrogen production. The composites can achieve a photocatalytic hydrogen production rate of up to 7.8 mmol·g^{-1}·h^{-1}, which is about 3.69 times higher than that of cadmium sulfide without biochar.

The regulation of photophysical properties is a key focus of research in luminescent materials. Xia's group [22] studied the effect of organic spacer cations on the optical properties of quasi-two-dimensional perovskite. An organic spacer with short chain length can greatly reduce the quantum confinement and dielectric confinement in perovskite. Considering the impact of synthetic conditions on the structural properties of materials, Lee's group [23] investigated their effect on the optical properties of Si-substituted CaYAlO$_4$:Eu. The use of ball milling can reduce the particle size and induce surface defects in Al$_2$O$_3$, and the ratio of the charge transfer band to f-f transition increases as the ball milling time increases. On the contrary, the variation in aluminum precursors has a negligible impact on the quantum efficiency of CaYAlO$_4$:Eu. Metal–organic coordination is also a powerful strategy for adjusting optical properties. Malandrinos's group [24] studied the influence of organic ligands on the optical properties of silver complexes. Different ligands can significantly regulate the quantum efficiency (11–23%) of silver complexes. Interestingly, the solid-state luminescence behavior of silver complexes is obviously different from that in a solution. In contrast, Urriolabeitia's group [25] explored the influence of Pd coordination on the photophysical properties of (Z)-4-hetarylidene-5(4H)-oxazolones complexes. Their results show that the coordination of Pd^{2+} ions with 4-hetaryliden-5(4H)-oxazolone does not cause, in these cases, an increase in fluorescence intensity.

Therefore, the articles collected in this Special Issue report on the most pressing issues in the field of luminescent materials, such as the development of advanced synthesis technologies, the construction of novel luminescent materials, and the regulation of photophysical properties.

Conflicts of Interest: The author declares no conflicts of interest.

References

1. Chen, G.L.; Feng, H.; Jiang, X.G.; Xu, J.; Pan, S.F.; Qian, Z.S. Redox-controlled fluorescent nanoswitch based on reversible disulfide and its application in butyrylcholinesterase activity assay. *Anal. Chem.* **2018**, *90*, 1643–1651. [CrossRef] [PubMed]
2. Lu, X.M.; Zhang, J.Y.; Xie, Y.N.; Zhang, X.F.; Jiang, X.M.; Hou, X.D.; Wu, P. Ratiometric phosphorescent probe for thallium in serum, water, and soil samples based on long-lived, spectrally resolved, Mn-doped ZnSe quantum dots and carbon dots. *Anal. Chem.* **2018**, *90*, 2939–2945. [CrossRef] [PubMed]
3. Chen, J.; Jiang, X.; Zhang, C.; MacKenzie, K.R.; Stossi, F.; Palzkill, T.; Wang, M.C.; Wang, J. Reversible reaction-based fluorescent probe for real-time imaging of glutathione dynamics in mitochondria. *ACS Sens.* **2017**, *2*, 1257–1261. [CrossRef] [PubMed]
4. Zhang, S.; Chen, T.-H.; Lee, H.-M.; Bi, J.; Ghosh, A.; Fang, M.; Qian, Z.; Xie, F.; Ainsley, J.; Christov, C.; et al. Luminescent probes for sensitive detection of pH changes in live cells through two near-infrared luminescence channels. *ACS Sens.* **2017**, *2*, 924–931. [CrossRef] [PubMed]
5. Gao, X.; Zhang, W.; Dong, Z.; Ren, J.; Song, B.; Zhang, R.; Yuan, J. FRET luminescent probe for the ratiometric imaging of peroxynitrite in rat brain models of epilepsy-based on organic dye-conjugated iridium(III) complex. *Anal. Chem.* **2023**, *95*, 18530–18539. [CrossRef]
6. Ma, Y.; Tang, Y.; Zhao, Y.; Lin, W. Rational design of a reversible fluorescent probe for sensing sulfur dioxide/formaldehyde in living cells, zebrafish, and living mice. *Anal. Chem.* **2019**, *91*, 10723–10730. [CrossRef]
7. Wu, W.T.; Zhan, L.Y.; Fan, W.Y.; Song, J.Z.; Li, X.M.; Li, Z.T.; Wang, R.Q.; Zhang, J.Q.; Zheng, J.T.; Wu, M.B. Cu-N dopants boost electron transfer and photooxidation reactions of carbon dots. *Angew. Chem. Int. Ed.* **2015**, *127*, 6640–6644. [CrossRef]
8. Lou, X.-Y.; Zhang, G.; Li, M.-H.; Yang, Y.-W. Macrocycle-strutted coordination microparticles for fluorescence-monitored photosensitization and substrate-selective photocatalytic degradation. *Nano Lett.* **2023**, *23*, 1961–1969. [CrossRef]
9. Yuan, Y.J.; Yang, S.H.; Wang, P.; Yang, Y.; Li, Z.J.; Chen, D.Q.; Yu, Z.T.; Zou, Z.G. Bandgap-tunable black phosphorus quantum dots: Visible-light-active photocatalysts. *Chem. Commun.* **2018**, *54*, 960–963. [CrossRef]
10. Liang, L.Y.; Chen, B.B.; Wang, Y.; Gao, Y.T.; Chang, S.; Liu, M.L.; Li, D.W. Inorganic salt recrystallization strategy for achieving ultralong room temperature phosphorescence through structural confinement and aluminized reconstruction. *J. Colloid Interface Sci.* **2023**, *649*, 445–455. [CrossRef]
11. Chen, B.-B.; Chang, S.; Lv, J.; Qian, R.-C.; Li, D.-W. Temperature-modulated porous gadolinium micro-networks with hyperchrome-enhanced fluorescence effect. *Chem. Eng. J.* **2021**, *422*, 129959. [CrossRef]

12. Satoh, C.; Okada, T.; Oono, T.; Sasaki, T.; Shimizu, T.; Fukagawa, H. Bandgap engineering for ultralow-voltage operation of organic light-emitting diodes. *Adv. Opt. Mater.* **2023**, *11*, 2300683. [CrossRef]
13. Schneider, E.M.; Bärtsch, A.; Stark, W.J.; Grass, R.N. Safe one-pot synthesis of fluorescent carbon quantum dots from lemon juice for a hands-on experience of nanotechnology. *J. Chem. Educ.* **2019**, *96*, 540–545. [CrossRef]
14. Sun, H.; Xia, P.; Shao, H.; Zhang, R.; Lu, C.; Xu, S.; Wang, C. Heating-free synthesis of red emissive carbon dots through separated processes of polymerization and carbonization. *J. Colloid Interface Sci.* **2023**, *646*, 932–939. [CrossRef] [PubMed]
15. Shao, Y.; Wang, Y.-L.; Tang, Z.; Wen, Z.; Chang, C.; Wang, C.; Sun, D.; Ye, Y.; Qiu, D.; Ke, Y.; et al. Scalable synthesis of photoluminescent single-chain nanoparticles by electrostatic-mediated intramolecular crosslinking. *Angew. Chem. Int. Ed.* **2022**, *61*, e202205183. [CrossRef]
16. Gao, Y.-T.; Chang, S.; Chen, B.-B.; Li, D.-W. Dual-exciting central carbon nanoclusters for the dual-channel detection of hemin. *Inorganics* **2023**, *11*, 226. [CrossRef]
17. Gorbachenya, K.N.; Volkova, E.A.; Maltsev, V.V.; Kisel, V.E.; Mitina, D.D.; Koporulina, E.V.; Kuzmin, N.N.; Marchenko, E.I.; Kosorukov, V.L. Growth, spectroscopic characterization and continuous-wave laser operation of Er,Yb:GdMgB$_5$O$_{10}$ crystal. *Inorganics* **2024**, *12*, 240. [CrossRef]
18. Anpalagan, K.; Yin, H.; Cole, I.; Zhang, T.; Lai, D.T.H. Quantum yield enhancement of carbon quantum dots using chemical-free precursors for sensing Cr (VI) ions. *Inorganics* **2024**, *12*, 96. [CrossRef]
19. Cheng, Y.; Huang, Y.; Yu, G. N-S-co-Doped carbon dot blue fluorescence preparation and baicalein detection. *Inorganics* **2024**, *12*, 154. [CrossRef]
20. Ishibashi, R.; Koike, R.; Suda, Y.; Kojima, T.; Sumi, T.; Misawa, T.; Kizu, K.; Okamura, Y.; Ito, T. Lanthanide-containing polyoxometalate crystallized with bolaamphiphile surfactants as inorganic–organic hybrid phosphors. *Inorganics* **2024**, *12*, 146. [CrossRef]
21. Li, X.; Zang, Y.; Zhang, J.; Zhang, L.; Zhang, J.; Huang, M.; Wang, J. Orange peel biochar–CdS composites for photocatalytic hydrogen production. *Inorganics* **2024**, *12*, 156. [CrossRef]
22. Zhang, L.; Xia, M.; Zhang, Y.; Song, L.; Guo, X.; Zhang, Y.; Wang, Y.; Xia, Y. The effect of organic spacer cations with different chain lengths on quasi-two-dimensional perovskite properties. *Inorganics* **2024**, *12*, 12. [CrossRef]
23. Oh, J.H.; Lee, Y.; Kim, J.; Hong, W.T.; Yang, H.K.; Kang, M.; Lee, S. Effect of synthesis conditions on the photoluminescent properties of Si-substituted CaYAlO$_4$:Eu: Sources of experimental errors in solid-state synthesis. *Inorganics* **2024**, *12*, 150. [CrossRef]
24. Glykos, D.; Tsipis, A.C.; Plakatouras, J.C.; Malandrinos, G. Synthesis, spectroscopic characterization, and photophysical studies of heteroleptic silver complexes bearing 2,9-bis(styryl)-1,10-phenanthroline ligands and bis[(2-diphenylphosphino)phenyl] ether. *Inorganics* **2024**, *12*, 131. [CrossRef]
25. Martínez, M.; Dalmau, D.; Crespo, O.; García-Orduña, P.; Lahoz, F.; Martín, A.; Urriolabeitia, E.P. Different patterns of Pd-promoted C-H bond activation in (Z)-4-hetarylidene-5(4H)-oxazolones and consequences in photophysical properties. *Inorganics* **2024**, *12*, 271. [CrossRef]

Disclaimer/Publisher's Note: The statements, opinions and data contained in all publications are solely those of the individual author(s) and contributor(s) and not of MDPI and/or the editor(s). MDPI and/or the editor(s) disclaim responsibility for any injury to people or property resulting from any ideas, methods, instructions or products referred to in the content.

Article

Dual-Exciting Central Carbon Nanoclusters for the Dual-Channel Detection of Hemin

Ya-Ting Gao [1], Shuai Chang [1], Bin-Bin Chen [1,2,*] and Da-Wei Li [1,*]

[1] Key Laboratory for Advanced Materials, Shanghai Key Laboratory of Functional Materials Chemistry, Frontiers Science Center for Materiobiology & Dynamic Chemistry, School of Chemistry & Molecular Engineering, East China University of Science and Technology, Shanghai 200237, China; gaoyating@mail.ecust.edu.cn (Y.-T.G.); y12213021@mail.ecust.edu.cn (S.C.)

[2] School of Science and Engineering, Shenzhen Institute of Aggregate Science and Technology, The Chinese University of Hong Kong, Shenzhen, 2001 Longxiang Boulevard, Longgang District, Shenzhen 518172, China

* Correspondence: chenbinbin_swu@163.com (B.-B.C.); daweili@ecust.edu.cn (D.-W.L.)

Abstract: Constructing optical nanoprobes with superior performance is highly desirable for sensitive and accurate assays. Herein, we develop a facile room-temperature strategy for the fabrication of green emissive carbon nanoclusters (CNCs) with dual-exciting centers for the dual-channel sensing of hemin. The formation of the CNCs is attributed to the crosslinking polymerization of the precursors driven by the Schiff base reaction between ethylenediamine and 2,3-dichloro-5,6-dicyano-1,4-benzoquinone. Most importantly, the proposed CNCs have a unique excitation-independent green emission (518 nm) with two excitation centers at 260 nm (channel 1) and 410 nm (channel 2). The dual-exciting central emission can serve as dual-channel fluorescence (FL) signals for highly sensitive and reliable detection of hemin based on the inner filter effect. Because of the great spectral overlap difference between the absorption spectrum of hemin and the excitation lights of the CNCs in the two channels, hemin has a different quenching effect on FL emission from different channels. The dual-channel signals of the CNCs can detect hemin in the range of 0.075–10 μM (channel 1) and 0.25–10 μM (channel 2), respectively. These findings not only offer new guidance for the facile synthesis of dual-exciting central CNCs but also establish a reliable sensing platform for the analysis of hemin in complex matrixes.

Keywords: carbon nanoclusters; dual-exciting centers; Schiff base reaction; dual-channel detection; hemin

Citation: Gao, Y.-T.; Chang, S.; Chen, B.-B.; Li, D.-W. Dual-Exciting Central Carbon Nanoclusters for the Dual-Channel Detection of Hemin. *Inorganics* **2023**, *11*, 226. https://doi.org/10.3390/inorganics11060226

Academic Editor: Zdeněk Slanina

Received: 19 April 2023
Revised: 18 May 2023
Accepted: 21 May 2023
Published: 25 May 2023

Copyright: © 2023 by the authors. Licensee MDPI, Basel, Switzerland. This article is an open access article distributed under the terms and conditions of the Creative Commons Attribution (CC BY) license (https://creativecommons.org/licenses/by/4.0/).

1. Introduction

Optical nanoprobes have attracted much attention because of their indispensable role in the field of sensing and imaging [1–3]. In addition to their high sensitivity and selectivity for analytes, excellent nanoprobes usually require a high signal-to-noise ratio to meet the rapidly increasing requirements of food safety, disease diagnosis, environmental analysis, and other fields. Single-signal-based nanoprobes, the most common sensing system, are usually problematic due to the unavoidable interferences during the detection of analytes, such as the non-homogeneous concentration distribution of nanoprobes, light scattering from the sample matrixes, and excitation light fluctuation [4]. In order to overcome these drawbacks, various strategies have been developed to achieve the accurate measurement of analytes, the most common of which is the design of ratiometric fluorescence (FL) and dual-channel sensing probes. Ratiometric FL nanoprobes are attractive for the improvement of signal-to-noise ratio and the accuracy of analysis due to the intrinsic built-in correction [5]. Likewise, dual-channel sensing nanoprobes use dual-channel signals as self-control, which is beneficial for achieving more accurate analysis results [6].

The dual-channel sensing method can combine the advantages of each method and uses the dual-channel signals for self-supervision of detection results, which can improve the selectivity and accuracy of nanoprobes [7]. Currently, various strategies have been developed for the preparation of nanoprobes with dual-channel response signals. Wang's group [6] developed a colorimetric/electrochemical sensing platform for the detection of aflatoxin B1, with improved analysis precision and reduced false-negative and false-positive rates. Liu's group [8] developed a paper-based sensing system for fluorescent and colorimetric dual-channel determination of foodborne pathogenic bacteria, with more reliable results. Li's group [9] constructed periodic Au@metal–organic framework nanoparticle arrays as dual-channel biosensors for detecting glucose based on the dual signal change of surface plasmon resonance and diffraction peaks. These strategies usually rely on the response signals of a hybrid nanosystem composed of two optical materials that have a complicated and time-consuming preparation process [7]. Moreover, the stability of the hybrid nanosystem is a major obstacle to the development of dual-channel nanoprobes. Therefore, developing a facile strategy for the preparation of dual-channel nanoprobes is very attractive.

Hemin, a well-known natural porphyrinatoiron complex, is an indispensable substance in living organisms, which has important roles in the regulation of oxygen transport, gene expression, hemoglobin synthesis, and other physiological processes [10–12]. Hemin is composed of a porphyrin ring containing four pyrrole molecules and Fe^{3+} ions located at the center, which can keep stable in the body as a cofactor of hemoglobin without resulting in hemolysis and inflammation [13]. Hemin is abundant in living organisms mainly as an electron transfer medium according to the reversible redox reaction of Fe^{3+}/Fe^{2+} [14]. Excessive or lack of hemin is not beneficial for human health; this deficiency may cause neuroglobin expression in neural cells and accelerate the formation of endogenous CO, while an excess may lead to permanent brain secondary damage after a hemorrhagic stroke [15]. In addition, hemin can not only serve as a natural iron supplement for the treatment of iron deficiency anemia but also acts as a raw material for the synthesis of semi-synthetic bilirubin and anticancer medicines [16]. Therefore, accurate determination of hemin is of great significance for the early diagnosis of diseases and analysis of medical content. Dual-channel sensing platforms provide a great potential for highly sensitive and reliable detection of hemin, which exhibits incomparable advantages such as more reliable analysis results in comparison to other traditional methods, including chemiluminescence [16], electrochemiluminescence [17], and surface-enhanced Raman scattering [18] techniques.

Luminescent carbon nanoclusters (CNCs) are regarded as a new type of carbon nanomaterial, which has been widely used in numerous fields, such as optical imaging and analysis detection [5]. In contrast to the widely studied single-exciting central CNCs, dual-exciting central CNCs have rarely been reported to date. Our group has developed various quinone-based luminescent nanocarbon materials with different emission properties [19–21], such as dual-emission carbonized polymer dots prepared by using ethylenediamine and tetrachlorobenzoquinone at room temperature [22], whereas CNCs with dual excitation centers have not been reported yet. By changing tetrachlorobenzoquinone to 2,3-dichloro-5,6-dicyano-1,4-benzoquinone (DDQ), for the first time, we developed a facile Schiff base reaction for the preparation of dual-exciting central CNCs by simply keeping the mixture of ethylenediamine (EDA) and DDQ in aqueous solution at room temperature (Scheme 1a). DDQ was chosen as a precursor because its active carbonyl group can react with the amino group of EDA at room temperature to form CNCs through Schiff base polymerization. The proposed CNCs exhibit a unique dual-exciting central emission property: one excitation-independent green emission, but can be excited at two different excitation regions centered at about 260 nm (channel 1) and 410 nm (channel 2), respectively. Because of the difference of spectral overlap, hemin has different FL quenching ability to the two channels of the CNCs (Scheme 1b), achieving a channel-dependent detection range towards hemin, with a linear range of 0.075–10 µM (channel 1) and 0.25–10 µM (channel 2), respectively. This work not only provides a deeper understanding of the optical mechanism of luminescent carbon materials but also develops a reliable dual-channel sensing platform for hemin detection.

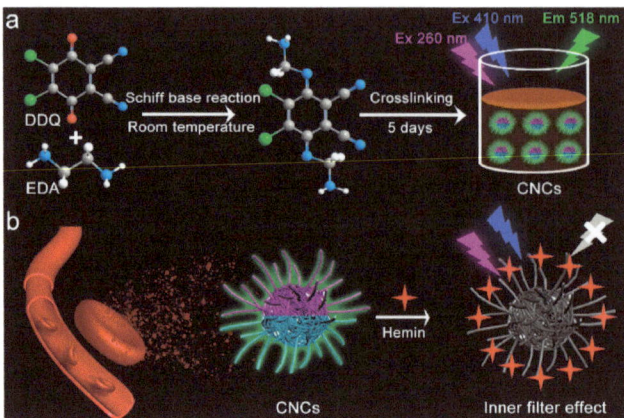

Scheme 1. Illustration of the room-temperature synthesis of dual-exciting central CNCs using EDA and DDQ and their dual-channel detection of hemin.

2. Results and Discussion

2.1. Synthesis and Characterization of CNCs

A facile room temperature method is developed for the first time for preparing dual-exciting central CNCs by mixing the EDA and DDQ in an aqueous solution. When EDA and DDQ are mixed together, two active amino groups of the EDA molecule can easily react with the two carbonyl groups of DDQ by Schiff base reaction to form an extended polymer chain [22]. Subsequently, intermolecular interactions enable the polymer chains to be further wound into amorphous CNCs [5]. A high-resolution transmission electron microscope (TEM) image (Figure 1a) exhibits that the proposed CNCs do not have an obvious carbon core or any lattice fringes, indicating their completely amorphous internal structure. The particle statistics result (Figure 1b) shows that the CNCs have a uniform size distribution with an average diameter of 1.8 ± 0.2 nm.

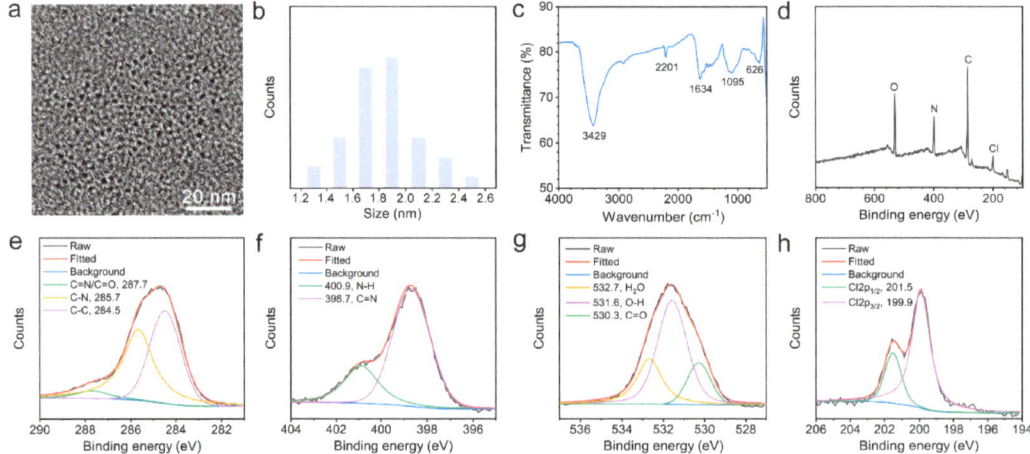

Figure 1. Characterization of dual-exciting central CNCs. (**a**) TEM image, (**b**) size distribution, (**c**) FT-IR spectrum, (**d**) XPS spectrum, (**e**) C1s spectrum, (**f**) N1s spectrum, (**g**) O1s spectrum and (**h**) Cl2p spectrum of CNCs.

The structural characteristics of the CNCs are further studied by Fourier transform infrared (FT-IR) and X-ray photoelectron spectrometer (XPS) spectra. FT-IR spectrum (Figure 1c) shows that the prepared CNCs have several strong stretching vibration bands of N–H at 3429 cm^{-1}, C≡N at 2201 cm^{-1}, C=N at 1634 cm^{-1} and C–Cl at 626 cm^{-1} [23]. The appearance of the C=N bond strongly confirms that the formation of the CNCs is attributed to the crosslinking polymerization of precursors driven by the Schiff base reaction. The XPS spectrum (Figure 1d) clearly reveals that the prepared CNCs are composed of carbon (C1s, 284.8 eV), nitrogen (N1s, 399.3 eV), oxygen (O1s, 531.2 eV), and chlorine (Cl2p, 200.5 eV) elements. The high-resolution C1s spectrum (Figure 1e) exhibits three peaks at 287.7, 285.7, and 284.5 eV, which are ascribed to C=O/C=N, C–N, and C–C bonds, respectively [24]. In addition, the high-resolution N1s spectrum (Figure 1f) presents two peaks at 398.7 and 400.9 eV, which are attributed to C=N and N–H bonds, respectively [25]. Moreover, three peaks at 532.7, 531.6, and 530.3 eV in the O1s spectrum (Figure 1g) can be attributed to the adsorbed H$_2$O, O–H, and C=O bonds, respectively [26,27]. Due to the spin-orbit splitting of the Cl2p core level, the Cl2p spectrum (Figure 1h) shows two peaks centered at 201.5 and 199.9 eV, which can be attributed to Cl2p$_{1/2}$ and Cl2p$_{3/2}$, respectively [28]. These results demonstrate that the prepared CNCs are a kind of ultra-small amorphous carbon nanoparticles containing rich C=N and C≡N groups.

2.2. Optical Properties of CNCs

The optical properties of CNCs are investigated in detail by absorption and FL spectra. As shown in Figure 2a, the absorption spectrum of CNCs exhibits two strong absorption bands at 257 and 407 nm, which are due to the $\pi-\pi^*$ transition of the conjugated sp^2 structure and $n-\pi^*$ transition of aggregated fluorescent groups [29]. Correspondingly, two excitation peaks have been found at 257 and 417 nm in the excitation spectrum of the CNCs. No matter if excited at 260 (channel 1) or 410 nm (channel 2), only one green emission peak of about 518 nm can be obtained, and there is only a diversity in intensity between the two. This unique dual-exciting central emission phenomenon is clearly displayed on a three-dimensional (3D) FL spectrum. The result (Figure 2b) shows that the CNCs show an excitation-independent green FL emission that can be excited at two excitation centers.

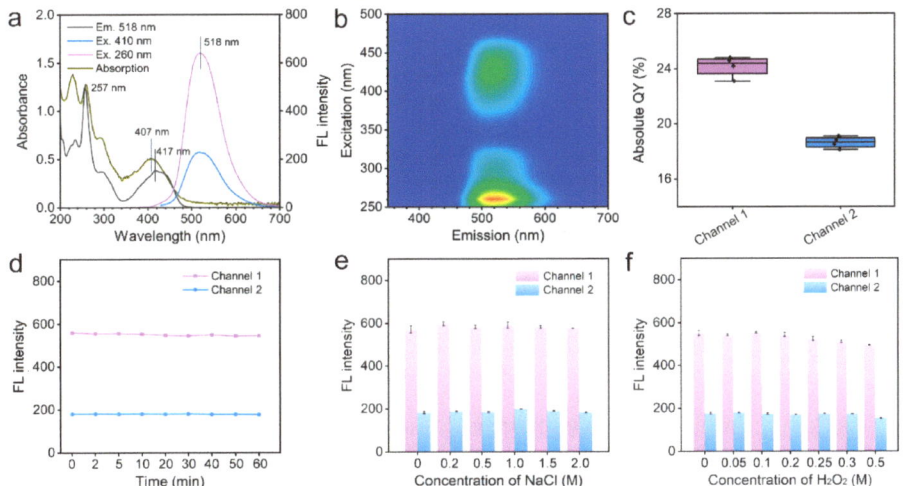

Figure 2. Optical properties and stability of dual-exciting central CNCs. (**a**) Absorption, excitation, and emission spectra, and (**b**) 3D FL spectrum of the CNCs. (**c**) Absolute QYs of the CNCs at different channels. (**d**) Photostability. (**e**) The stability in a salty medium, and (**f**) the antioxidant capacity of the CNCs.

Because of the lack of crystal structure, the strong green FL of the CNCs should be attributed to their surface state, rather than the quantum confinement effect [30]. The excitation peak of 257 nm may be ascribed to the $\pi-\pi^*$ transition of the conjugated sp^2 structure, while the excitation peak of 417 nm may be devoted to the $n-\pi^*$ transition of N-related groups such as the C = N bond. The dual-exciting central emission feature may be due to the hyperconjugation of their two FL centers [31]. Meanwhile, the luminescent efficiency of the CNCs is excitation-dependent, and the absolute quantum yields (QYs) of the CNCs are about 24.2% for channel 1 and 18.6% for channel 2 (Figure 2c).

Stability is an important parameter of nanoprobes, which determines the accuracy of the sensing system [32]. The traditional dye molecules are usually photosensitive and can easily cause FL quenching under strong light irradiation. On the contrary, the proposed CNCs exhibit good photostability, and their FL intensities at two excitations remain unchanged under 60 min of UV irradiation (Figure 2d), suggesting that they can be used for long-term imaging and sensing. Meanwhile, the FL intensities of the CNCs are also almost constant in a high ionic strength medium of up to 2 M (Figure 2e). Additionally, the proposed CNCs possess good antioxidant ability, and their FL emissions are not affected by the high concentration of H_2O_2 solution (Figure 2f). The solution pH has an important effect on the luminescence intensity of the CNCs. The result (Figure S1) shows that the FL intensities of CNCs at two excitations increase gradually with the increase in pH value from 1 to 11, which is usually due to the protonation and deprotonation of surface groups of CNCs controlled by the solution pH [21]. This result also indirectly indicates that the FL emission of the CNCs comes from the surface state.

2.3. Dual-Channel Fluorescent Detection of Hemin

Hemin can effectively quench the FL of the CNCs, which shows a channel-dependent quenching effect. The quenching process levels off after 1 min with the extension of the incubation time (Figure S2), suggesting that the sensing process is rapid. By using channel 1 (260 nm excitation), the ratio of the FL intensity of the CNCs is linearly correlated with the concentration of hemin in the range of 0.075–10 µM, with a low limit of detection of about 30 nM (Figure 3a,b). Common interferents, such as Na^+, Co^{2+}, L-glutathione (GSH), and cysteine (Cys) have little effect on the FL emission of CNCs in the absence and presence of hemin (Figure 3c), indicating that these interferents do not affect the detection of hemin by channel 1. By using channel 2 (410 nm excitation), the FL intensity ratio of the CNCs is proportional to the hemin concentration in the range of 0.25–10 µM with a limit of detection of about 90 nM (Figure 3d,e). Likewise, these interferents also do not affect the detection of hemin by channel 2 (Figure 3f). In view of this, the proposed CNCs can serve as dual-channel sensing nanoprobes for the detection of hemin with good selectivity. Other methods for detecting hemin usually rely on a single signal (Table S1). Although the limit of detection of the proposed probes is not the lowest, the high selectivity and accurate detection results with self-supervision make them powerful sensing probes for the detection of hemin.

Subsequently, the quenching effect of dual-exciting central CNCs by hemin is explored. The inner filter effect is regarded as a non-radiative energy conversion process, which can be well used for the detection of analytes by competitive light absorption. A highly efficient inner filter effect usually requires a great spectral overlap between the absorption band of analytes and the excitation/emission band of nanoprobes [33]. The high selectivity of hemin detection is attributed to the fact that the absorption band of common interfering substances, such as GSH and Cys, is usually located in the deep ultraviolet region, which has no effect on the excitation and emission light of the CNCs and, therefore, does not interfere with the detection of hemin. As shown in Figure 4a, the absorption band of hemin has a greater overlap with the excitation band of channel 1 of CNCs in comparison to the excitation band of channel 2 of CNCs. This result shows that hemin can more effectively absorb the excitation light of channel 1 and achieve a more sensitive response. Moreover, no new absorption bands occur for the mixtures of CNCs and hemin, indicating that there is no formation of new complexes between the CNCs and hemin (Figure 4b). Meanwhile, the FL decay curves (Figure 4c,d) of CNCs exhibit that there are almost no variations in a lifetime for CNCs in the absence and presence of hemin. These results strongly confirm that the FL quenching of the CNCs induced by hemin is due to the inner filter effect.

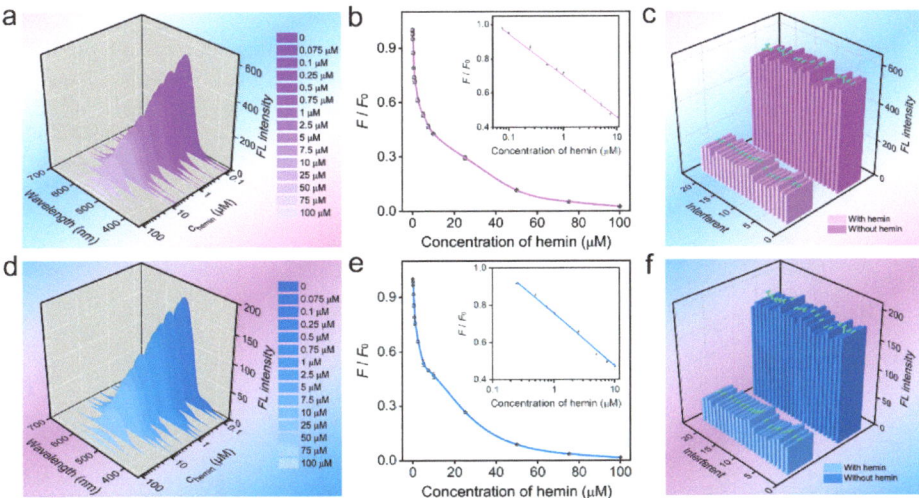

Figure 3. Sensitivity and selectivity of hemin detection. The FL spectra of CNCs in the presence of hemin excited at (**a**) 260 nm (channel 1) and (**d**) 410 nm (channel 2). Sensitivity of hemin detection by (**b**) channel 1 and (**e**) channel 2. FL responses of CNCs in the presence of 50 µM interferents by (**c**) channel 1 and (**f**) channel 2. The columns represent the following: 1, control; 2, Na^+; 3, Mg^{2+}; 4, K^+; 5, Co^{2+}; 6, Ni^{2+}; 7, Cu^{2+}; 8, Ag^+; 9, Cd^{2+}; 10, Ba^{2+}; 11, Pb^{2+}; 12, Cl^-; 13, Pb^{2+}; 14, HCO_3^-; 15, $H_2PO_4^-$; 16, Lys; 17, Thr; 18, Val; 19, glucose; 20, dopamine; 21, Cys; 22, Hcy; 23, GSH. Concentration of hemin: 25 µM. Error bars represent ± S.D (n = 3).

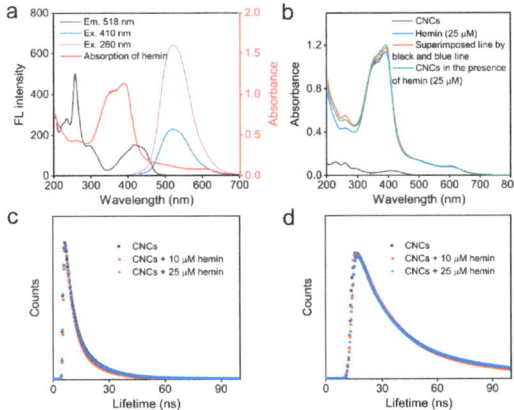

Figure 4. The sensing mechanism of hemin using CNCs. (**a**) Overlap degree of the absorption band of hemin and excitation and emission bands of CNCs at two channels. (**b**) The absorption spectra of CNCs in the absence and presence of hemin. (**c**) Time-resolved decay of CNCs before and after adding hemin (10 mM or 25 mM). Excitation: 260 nm. (**d**) Time-resolved decay of CNCs before and after adding hemin (10 mM or 25 mM). Excitation: 410 nm.

2.4. Determination of Hemin in Cells

Before cell imaging, the cytotoxicity of CNCs is tested. The cytotoxicity of CNCs is relatively low based on the result of the CKK-8 assay assessment (Figure S3), and over 90% HeLa cell viability is observed after 24 h incubation with the CNCs (0.05–5 mg/mL). The real-time monitoring of the dynamic invasion process of hemin into HeLa cells is further investigated. As shown in Figure 5, HeLa cells display bright green FL after incubation with CNCs for 2 h, indicating that CNCs efficiently enter into the cells by endocytosis. Some reports reveal that both caveolae- and clathrin-mediated endocytosis is the main cellular uptake mechanism of nanocarbon [34], indicating that multiple endocytosis pathways may cause the high transfection efficiency of CNCs. After the addition of hemin for 30 min, the green FL of CNCs is effectively quenched, and the degree of quenching is closely related to the concentration of hemin. Therefore, the proposed CNCs can be used as fluorescent nanoprobes to monitor the dynamic intrusion process of hemin into cells.

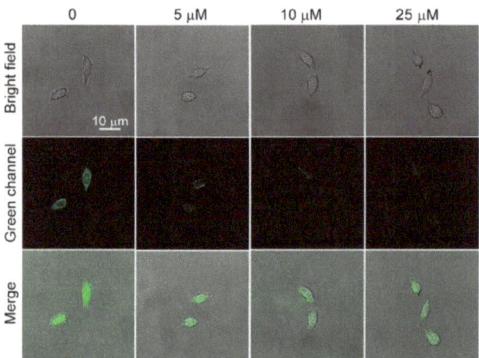

Figure 5. Laser confocal microscopic images of HeLa cells incubated with different concentrations of hemin. Excitation: 405 nm. Incubation time: 30 min. Concentration of CNCs: 1 mg/mL.

3. Materials and Methods

3.1. Materials and Reagents

Dopamine, GSH, Cys, homocysteine (Hcy), DDQ, and EDA are obtained from Aladdin Reagent Co., Ltd. (Shanghai, China). Glucose, threonine (Thr), lysine (Lys), and valine (Val) are from Macklin Biochemical Co., Ltd. (Shanghai, China). All reagents are used without any further purification, which are dissolved in 18.2 MΩ.cm ultrapure water.

3.2. Apparatus and Characterization

The morphology of CNCs is obtained by a Talos F200XTEM. The FL spectra of CNCs are measured by an F97Pro FL spectrophotometer. The absorption spectra of CNCs are measured by a 759S UV-Vis spectrophotometer. The FL lifetimes of CNCs are obtained by an FLS1000 FL spectrometer. The surface element composition of CNCs is determined using an XPS.

3.3. Preparation and Purification of CNCs

The synthesis of CNCs is achieved by the Schiff base reaction between DDQ and EDA. Briefly, 10 mg DDQ and 0.3 mL EDA are added to 4.7 mL ultrapure water in turn. The CNCs can be formed after 5 days of room-temperature reaction. The proposed CNCs are dialyzed using a 100–500 MWCO dialysis membrane for 2 days.

3.4. Sensing Procedure of Hemin Using CNCs

CNCs solution (20 μL) is firstly added to 1.78 mL water, and then different concentrations of hemin (0.2 mL) are further added. The above solution is incubated for 2 min and then transferred to a quartz cell for FL measurement. The selectivity of CNCs nanoprobes is studied by adding various interferents such as ions (Na^+, Mg^{2+}, K^+, Co^{2+}, Ni^{2+}, Cu^{2+}, Ag^+, Cd^{2+}, Ba^{2+}, Pb^{2+}, Cl^-, HCO_3^-, and $H_2PO_4^-$), amino acids (Lys, Thr, and Val), and other substances (GSH, Cys, Hcy, dopamine, and glucose) that are used to replace hemin at a high concentration of 50 μM. The excitation light is selected at 260 and 410 nm as the dual-channel signals for detection.

4. Conclusions

In conclusion, a facile and one-pot Schiff base reaction is developed for the fabrication of dual-exciting central CNCs at room temperature. The excitation-independent green emission of CNCs can be excited at two different excitation regions (260 nm and 410 nm), showing the unique optical phenomenon of dual-excitation and single-emission. Additionally, the proposed CNCs with low toxicity exhibit good oxidation and photobleaching resistance. Based on the dual-exciting emission properties, CNCs can serve as dual-channel FL nanoprobes for the highly sensitive detection of hemin based on the inner filter effect. Meanwhile, CNCs display high reproducibility for sensing hemin in living cells, providing new insights to design a reliable dual-channel sensing nanoprobe for analytical and biomedical applications.

Supplementary Materials: The following supporting information can be downloaded at: https://www.mdpi.com/article/10.3390/inorganics11060226/s1, Figure S1: effect of solution pH on FL emissions of CNCs at two channels; Figure S2: relationship between FL intensity and incubation time; Table S1: the comparison of the determination of hemin [35,36]; Figure S3: the biocompatibility of the CNCs.

Author Contributions: Data curation, B.-B.C.; formal analysis, Y.-T.G. and S.C.; funding acquisition, D.-W.L.; methodology, Y.-T.G.; project administration, D.-W.L.; software, Y.-T.G.; supervision, B.-B.C. and D.-W.L.; validation, Y.-T.G.; visualization, B.-B.C.; writing—original draft, B.-B.C.; writing—review and editing, B.-B.C. and D.-W.L. All authors have read and agreed to the published version of the manuscript.

Funding: The authors appreciate the financial support from the National Natural Science Foundation of China (21974046, 22176058, and 21977031), the Science and Technology Commission of Shanghai Municipality (19520744000 and 19ZR1472300), and the Fundamental Research Funds for the Central Universities (222201717003).

Data Availability Statement: Not appliance.

Acknowledgments: The authors thank the Research Center of Analysis and Test of East China University of Science and Technology for the help with the characterization.

Conflicts of Interest: The authors declare no conflict of interest.

References

1. Liu, M.L.; Chen, B.B.; Li, C.M.; Huang, C.Z. Carbon dots prepared for fluorescence and chemiluminescence sensing. *Sci. China Chem.* **2019**, *62*, 968–981. [CrossRef]
2. Chen, B.B.; Liu, M.L.; Huang, C.Z. Recent advances of carbon dots in imaging-guided theranostics. *TrAC Trends Anal. Chem.* **2021**, *134*, 116116. [CrossRef]
3. Park, S.-H.; Kwon, N.; Lee, J.-H.; Yoon, J.; Shin, I. Synthetic ratiometric fluorescent probes for detection of ions. *Chem. Soc. Rev.* **2020**, *49*, 143–179. [CrossRef]
4. Gui, R.; Jin, H.; Bu, X.; Fu, Y.; Wang, Z.; Liu, Q. Recent advances in dual-emission ratiometric fluorescence probes for chemo/biosensing and bioimaging of biomarkers. *Coordin. Chem. Rev.* **2019**, *383*, 82–103. [CrossRef]
5. Chen, B.-B.; Liu, M.-L.; Gao, Y.-T.; Chang, S.; Qian, R.-C.; Li, D.-W. Design and applications of carbon dots-based ratiometric fluorescent probes: A review. *Nano Res.* **2023**, *16*, 1064–1083. [CrossRef]
6. Qian, J.; Ren, C.; Wang, C.; An, K.; Cui, H.; Hao, N.; Wang, K. Gold nanoparticles mediated designing of versatile aptasensor for colorimetric/electrochemical dual-channel detection of aflatoxin B1. *Biosens. Bioelectron.* **2020**, *166*, 112443. [CrossRef] [PubMed]
7. Li, W.; Zhang, X.; Hu, X.; Shi, Y.; Liang, N.; Huang, X.; Wang, X.; Shen, T.; Zou, X.; Shi, J. Simple design concept for dual-channel detection of Ochratoxin A based on bifunctional metal–organic framework. *ACS Appl. Mater. Interfaces* **2022**, *14*, 5615–5623. [CrossRef]

8. Wang, C.; Gao, X.; Wang, S.; Liu, Y. A smartphone-integrated paper sensing system for fluorescent and colorimetric dual-channel detection of foodborne pathogenic bacteria. *Anal. Bioanal. Chem.* **2020**, *412*, 611–620. [CrossRef]
9. Hang, L.; Zhou, F.; Men, D.; Li, H.; Li, X.; Zhang, H.; Liu, G.; Cai, W.; Li, C.; Li, Y. Functionalized periodic Au@MOFs nanoparticle arrays as biosensors for dual-channel detection through the complementary effect of SPR and diffraction peaks. *Nano Res.* **2017**, *10*, 2257–2270. [CrossRef]
10. Fereja, S.L.; Fang, Z.; Li, P.; Guo, J.; Fereja, T.H.; Chen, W. "Turn-off" sensing probe based on fluorescent gold nanoclusters for the sensitive detection of hemin. *Anal. Bioanal. Chem.* **2021**, *413*, 1639–1649. [CrossRef]
11. Gao, L.; Xiao, Y.; Wang, Y.; Chen, X.; Zhou, B.; Yang, X. A carboxylated graphene and aptamer nanocomposite-based aptasensor for sensitive and specific detection of hemin. *Talanta* **2015**, *132*, 215–221. [CrossRef] [PubMed]
12. Gao, S.; Wang, R.; Bi, Y.; Qu, H.; Chen, Y.; Zheng, L. Identification of frozen/thawed beef based on label-free detection of hemin (Iron Porphyrin) with solution-gated graphene transistor sensors. *Sens. Actuators B Chem.* **2020**, *305*, 127167. [CrossRef]
13. Du, N.; Zhang, H.; Wang, J.; Dong, X.; Li, J.; Wang, K.; Guan, R. Fluorescent silicon nanoparticle–based quantitative hemin assay. *Anal. Bioanal. Chem.* **2022**, *414*, 8223–8232. [CrossRef] [PubMed]
14. Zhao, L.; Chen, F.; Huang, W.; Bao, H.; Hu, Y.; Huang, X.-a.; Deng, T.; Liu, F. A fluorescence turn-on assay for simple and sensitive determination of hemin and blood stains. *Sens. Actuators B Chem.* **2020**, *304*, 127392. [CrossRef]
15. Ni, P.; Chen, C.; Jiang, Y.; Lu, Y.; Chen, W. A simple and sensitive fluorescent assay for hemin detection based on artemisinin-thiamine. *Sens. Actuators B Chem.* **2018**, *273*, 198–203. [CrossRef]
16. Fereja, T.H.; Kitte, S.A.; Gao, W.; Yuan, F.; Snizhko, D.; Qi, L.; Nsabimana, A.; Liu, Z.; Xu, G. Artesunate-luminol chemiluminescence system for the detection of hemin. *Talanta* **2019**, *204*, 379–385. [CrossRef]
17. Bushira, F.A.; Kitte, S.A.; Xu, C.; Li, H.; Zheng, L.; Wang, P.; Jin, Y. Two-dimensional-plasmon-boosted iron single-atom electrochemiluminescence for the ultrasensitive detection of dopamine, hemin, and mercury. *Anal. Chem.* **2021**, *93*, 9949–9957. [CrossRef]
18. Li, D.; Li, C.; Liang, A.; Jiang, Z. SERS and fluorescence dual-mode sensing trace hemin and K+ based on G-quarplex/hemin DNAzyme catalytic amplification. *Sens. Actuators B Chem.* **2019**, *297*, 126799. [CrossRef]
19. Liu, M.L.; Yang, L.; Li, R.S.; Chen, B.B.; Liu, H.; Huang, C.Z. Large-scale simultaneous synthesis of highly photoluminescent green amorphous carbon nanodots and yellow crystalline graphene quantum dots at room temperature. *Green Chem.* **2017**, *19*, 3611–3617. [CrossRef]
20. Chang, S.; Chen, B.B.; Lv, J.; Fodjo, E.K.; Qian, R.C.; Li, D.W. Label-free chlorine and nitrogen-doped fluorescent carbon dots for target imaging of lysosomes in living cells. *Microchim. Acta* **2020**, *187*, 435–442. [CrossRef]
21. Chen, B.B.; Liu, Z.X.; Deng, W.C.; Zhan, L.; Liu, M.L.; Huang, C.Z. A large-scale synthesis of photoluminescent carbon quantum dots: A self-exothermic reaction driving the formation of the nanocrystalline core at room temperature. *Green Chem.* **2016**, *18*, 5127–5132. [CrossRef]
22. Gao, Y.-T.; Chen, B.-B.; Jiang, L.; Lv, J.; Chang, S.; Wang, Y.; Qian, R.-C.; Li, D.-W.; Hafez, M.E. Dual-emitting carbonized polymer dots synthesized at room temperature for ratiometric fluorescence sensing of vitamin B12. *ACS Appl. Mater. Interfaces* **2021**, *13*, 50228–50235. [CrossRef]
23. Marković, Z.M.; Labudová, M.; Danko, M.; Matijašević, D.; Mičušík, M.; Nádaždy, V.; Kováčová, M.; Kleinová, A.; Špitalský, Z.; Pavlović, V.; et al. Highly efficient antioxidant F- and Cl-doped carbon quantum dots for bioimaging. *ACS Sustain. Chem. Eng.* **2020**, *8*, 16327–16338. [CrossRef]
24. Liu, M.L.; Chen, B.B.; He, J.H.; Li, C.M.; Li, Y.F.; Huang, C.Z. Anthrax biomarker: An ultrasensitive fluorescent ratiometry of dipicolinic acid by using terbium(III)-modified carbon dots. *Talanta* **2019**, *191*, 443–448. [CrossRef]
25. Stevens, J.S.; de Luca, A.C.; Pelendritis, M.; Terenghi, G.; Downes, S.; Schroeder, S.L.M. Quantitative analysis of complex amino acids and RGD peptides by X-ray photoelectron spectroscopy (XPS). *Surf. Interface Anal.* **2013**, *45*, 1238–1246. [CrossRef]
26. Kloprogge, J.T.; Duong, L.V.; Wood, B.J.; Frost, R.L. XPS study of the major minerals in bauxite: Gibbsite, bayerite and (pseudo-)boehmite. *J. Colloid Interface Sci.* **2006**, *296*, 572–576. [CrossRef]
27. Yang, D.; Velamakanni, A.; Bozoklu, G.; Park, S.; Stoller, M.; Piner, R.D.; Stankovich, S.; Jung, I.; Field, D.A.; Ventrice, C.A., Jr. Chemical analysis of graphene oxide films after heat and chemical treatments by X-ray photoelectron and Micro-Raman spectroscopy. *Carbon* **2009**, *47*, 145–152. [CrossRef]
28. Gu, J.; Hu, M.J.; Guo, Q.Q.; Ding, Z.F.; Sun, X.L.; Yang, J. High-yield synthesis of graphene quantum dots with strong green photoluminescence. *RSC Adv.* **2014**, *4*, 50141–50144. [CrossRef]
29. Bai, L.; Yan, H.; Feng, Y.; Feng, W.; Yuan, L. Multi-excitation and single color emission carbon dots doped with silicon and nitrogen: Synthesis, emission mechanism, Fe^{3+} probe and cell imaging. *Chem. Eng. J.* **2019**, *373*, 963–972. [CrossRef]
30. Liu, M.L.; Chen, B.B.; Li, C.M.; Huang, C.Z. Carbon dots: Synthesis, formation mechanism, fluorescence origin and sensing applications. *Green Chem.* **2019**, *21*, 449–471. [CrossRef]
31. Liao, X.; Chen, C.; Yang, J.; Zhou, R.; Si, L.; Huang, Q.; Huang, Z.; Lv, C. Nitrogen-doped carbon dots for dual-wavelength excitation fluorimetric assay for ratiometric determination of phosalone. *Microchim. Acta* **2021**, *188*, 247. [CrossRef] [PubMed]
32. Chen, B.B.; Liu, H.; Huang, C.Z.; Ling, J.; Wang, J. Rapid and convenient synthesis of stable silver nanoparticles with kiwi juice and its novel application for detecting protease K. *New J. Chem.* **2015**, *39*, 1295–1300. [CrossRef]
33. Chen, S.; Yu, Y.L.; Wang, J.H. Inner filter effect-based fluorescent sensing systems: A review. *Anal. Chim. Acta* **2018**, *999*, 13–26. [CrossRef]

34. Cao, X.; Wang, J.P.; Deng, W.W.; Chen, J.J.; Wang, Y.; Zhou, J.; Du, P.; Xu, W.Q.; Wang, Q.; Wang, Q.L.; et al. Photoluminescent cationic carbon dots as efficient non-viral delivery of plasmid SOX9 and chondrogenesis of fibroblasts. *Sci. Rep.* **2018**, *8*, 7057. [CrossRef] [PubMed]
35. Kang, B.H.; Li, N.; Liu, S.G.; Li, N.B.; Luo, H.Q. A label-free, highly sensitive and selective detection of hemin based on the competition between hemin and protoporphyrin IX binding to G-quadruplexes. *Anal. Sci.* **2016**, *32*, 887–892. [CrossRef] [PubMed]
36. Guo, Z.; Li, B.; Zhang, Y.; Zhao, Q.; Zhao, J.; Li, L.; Zuo, G. Acid-treated Graphitic Carbon Nitride Nanosheets as Fluorescence Probe for Detection of Hemin. *ChemistrySelect* **2019**, *4*, 8178–8182. [CrossRef]

Disclaimer/Publisher's Note: The statements, opinions and data contained in all publications are solely those of the individual author(s) and contributor(s) and not of MDPI and/or the editor(s). MDPI and/or the editor(s) disclaim responsibility for any injury to people or property resulting from any ideas, methods, instructions or products referred to in the content.

Article

N-S-co-Doped Carbon Dot Blue Fluorescence Preparation and Baicalein Detection

Yujia Cheng [1], Yan Huang [2] and Guang Yu [1,*]

[1] Mechanical and Electrical Engineering Institute, University of Electronic Science and Technology of China, Zhongshan Institute, Zhongshan 528400, China; chengyujia@zsc.edu.cn
[2] School of Materials and Energy, University of Electronic Science and Technology of China, Chengdu 610000, China; hyansworld@163.com
* Correspondence: yuguang@zsc.edu.cn; Tel.: +86-0760-8826-9835

Abstract: Carbon dots (CDs) have emerged as significant fluorescent nanomaterials due to their bright, stable fluorescence, good biocompatibility, facile synthesis, etc. They are widely used in various scientific and practical applications, particularly in combination with mesoporous, florescent, or magnetic nanomaterials to enhance their properties. Recent research has focused on employing CDs and their composites in drug analysis, drug loading, biological imaging, disease diagnosis, and temperature sensing, with a growing interest in their biological and medical applications. In this study, we synthesized blue-fluorescent S, N-co-doped CDs (cys-CDs) using hydrothermal synthesis with L-cysteine and sodium citrate. These resulting cys-CD particles were approximately 3.8 nm in size and exhibited stable fluorescence with a quantum yield of 0.66. By leveraging the fluorescence quenching of the cys-CDs, we developed a rapid and sensitive method for baicalein detection, achieving high sensitivity in the low micromolar range with a detection limit for baicalein of 33 nM. Our investigation revealed that the fluorescence-quenching mechanism involved static quenching and inner-filter effect components. Overall, cys-CDs proved to be effective for accurate quantitative baicalein detection in real-world samples.

Keywords: baicalein; carbon dots; fluorescence quenching; static quenching; inner-filter effect

Citation: Cheng, Y.; Huang, Y.; Yu, G. N-S-co-Doped Carbon Dot Blue Fluorescence Preparation and Baicalein Detection. *Inorganics* **2024**, *12*, 154. https://doi.org/10.3390/inorganics12060154

Academic Editors: Duncan H. Gregory and Binbin Chen

Received: 30 March 2024
Revised: 24 April 2024
Accepted: 30 April 2024
Published: 31 May 2024

Copyright: © 2024 by the authors. Licensee MDPI, Basel, Switzerland. This article is an open access article distributed under the terms and conditions of the Creative Commons Attribution (CC BY) license (https:// creativecommons.org/licenses/by/ 4.0/).

1. Introduction

Carbon dots (CDs) are a type of nanomaterial known for their advantageous fluorescent characteristics. These materials boast exceptional fluorescence performance, featuring bright, stable emission within narrow bandwidths, alongside notable biocompatibility and straightforward modification and functionalization [1–3]. Their utility spans various domains including drug testing, sensing biological small molecules, detecting ions, biological imaging, diagnosing diseases, and fabricating electro-optical devices. The research involving CDs has become increasingly diverse, with one notable focus being the enhancement of their fluorescence quantum yield (FQY) [4,5]. Heteroatom doping has emerged as a promising strategy for optimizing the fluorescence properties of CDs, proving to be an effective means to increase their FQY.

Baicalein, a flavonoid compound ($C_{15}H_{10}O_5$) with antibacterial and antiviral properties, is derived from the roots of *Scutellaria baicalensis* (*S. baicalensis*). Its chemical structure is illustrated in Figure 1. *S. baicalensis* can contain up to 5.41% baicalein. Given its potential for clinical applications, extensive research has been conducted on its structure, properties, modes of pharmacological mechanisms, and clinical uses. Several quantitative detection methods for baicalein have also been developed [6–8].

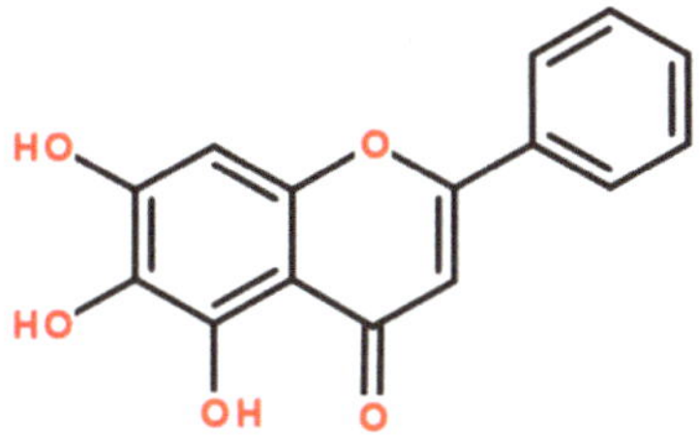

Figure 1. Chemical structure of baicalein.

Currently used detection methods, such as capillary electrophoresis–electrochemical analysis, high-performance liquid chromatography (HPLC), liquid chromatography-mass spectroscopy (LC-MS) detection, and liquid–liquid micro-extraction, involve complex sample preparation, resulting in high costs and low throughput [9]. Conversely, fluorescence detection offers numerous advantages, including real-time response, direct visualization and detection, low cost, the ease of operation, and the ability to develop high-throughput methodologies [10]. As a result, fluorescence detection is widely employed in various detection schemes.

In this study, we produced S, N-co-doped cysteine-modified carbon dots (cys-CDs) through hydrothermal synthesis, utilizing L-cysteine and sodium citrate as starting materials [11,12]. To analyze the particle size distribution, morphology, structure, composition, and fluorescence properties of the CDs, we employed techniques like infrared (IR) spectroscopy, UV-Vis absorption spectroscopy, transmission electron microscopy (TEM), X-ray photoelectron spectroscopy (XPS), and fluorescence measurements [13–15].

Subsequently, we investigated the fluorescence quenching of cys-CDs as a sensitive method for detecting baicalein. To validate the efficacy of our approach in real-world scenarios, we tested human blood and urine samples spiked with known concentrations of baicalein. Our results demonstrate that the detection of baicalein using cys-CDs is both sensitive and accurate in complex biological matrices. The overall methodology is illustrated in Figure 2.

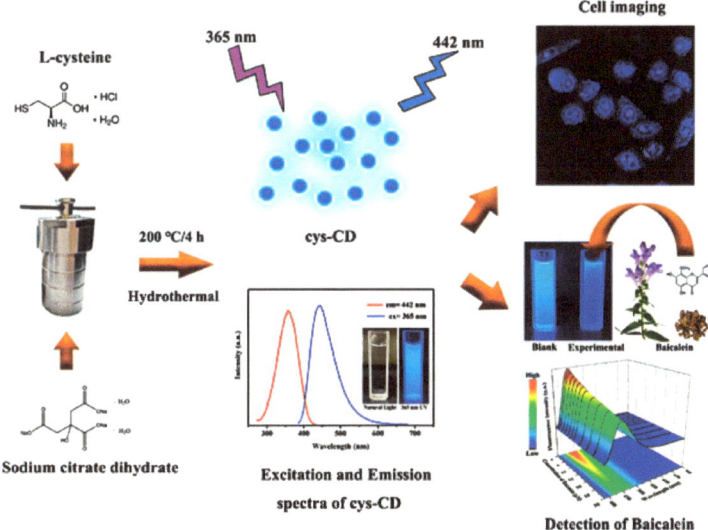

Figure 2. Hydrothermal synthesis of blue-fluorescent cys-CDs for baicalein detection and cell imaging.

2. Results and Discussion
2.1. Morphology and Structure of cys-CD

We employed TEM to analyze the morphological features, particle size distribution, and polydispersity of cys-CD (Figure 3a,b). The cys-CD particles appeared quasi-spherical, with an average diameter of 3.82 nm, ranging from 2.7 to 4.9 nm. Notably, the particle size distribution exhibited uniformity (Figure 3b).

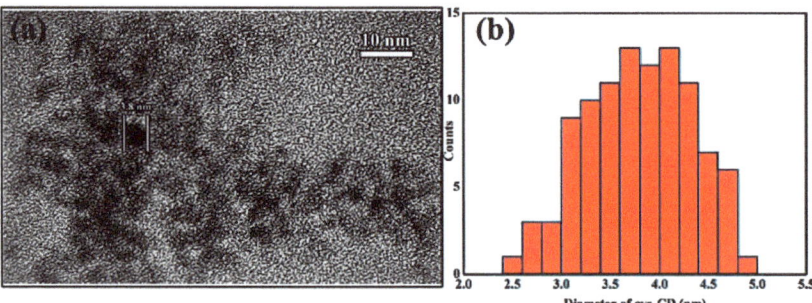

Figure 3. (a) Transmission electron microscopic image of cys-CD. (b) Particle size distribution of cys-CD.

Fourier-transform infrared (FT-IR) spectroscopy and XPS were employed to analyze the surface groups and chemical composition of cys-CDs [16]. In the IR spectrum (Figure 4a), we observed broad peaks at 3420 cm^{-1} and 3125 cm^{-1}, corresponding to O–H and N–H stretching vibration absorptions, respectively. Characteristic absorption peaks for C–N and –SH were detected around 2400 cm^{-1}, while the C=O stretching vibration peak appeared at 1660 cm^{-1}. The C–N stretching vibration absorption peak was observed at 1401 cm^{-1}, along with the C–S characteristic absorption peak at 1293 cm^{-1}. These observations indicate that the particle's surface maintained certain characteristics of the source material, showcasing abundant functional groups like amidogen, hydroxyl, carbonyl, and carboxyl groups. Compared with citric acid-CDs [17,18], this offers opportunities for further customization and the adjustment of the surface properties to suit specific applications.

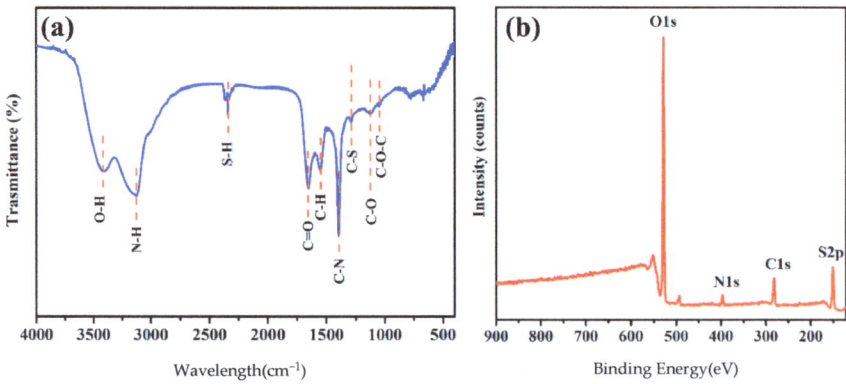

Figure 4. (a) FT-IR spectrum of cys-CD, (b) X-ray photoelectron spectrum of cys-CD.

The result of the XPS analysis of cys-CD is presented in Figure 4b. The binding energies at 398, 283, and 152 eV corresponded to N1s, C1s, and S2p, respectively, indicating the presence of carbon (C), nitrogen (N), and sulfur (S) and confirming the synthesis of the S, N-co-doped CDs. According to the research on the deconvolutions of the XPS peaks, the major forms of nitrogen elements are pyridine and pyrrole. The major forms of organic sulfur are sulfoxide and thioether [19].

2.2. Optical Properties of cys-CD

In natural light, a 100 μg mL^{-1} aqueous solution of cys-CD appeared as a faint yellow, transparent color. However, when exposed to ultraviolet (UV) light at 365 nm, the aqueous cys-CD solution emitted a vibrant blue fluorescence (see inset Figure 5a). This fluorescence is attributed to a π–π* transition occurring within the carbon dots, specifically from the C=O to the C=N groups. The peak excitation and emission wavelengths were recorded as λ_{ex} = 365 nm and λ_{em} = 442 nm, respectively (Figure 5a) [20]. Remarkably, we observed that the FQY of cys-CD reached up to 66%, surpassing the performance of the most blue-fluorescent CDs that have been reported previously.

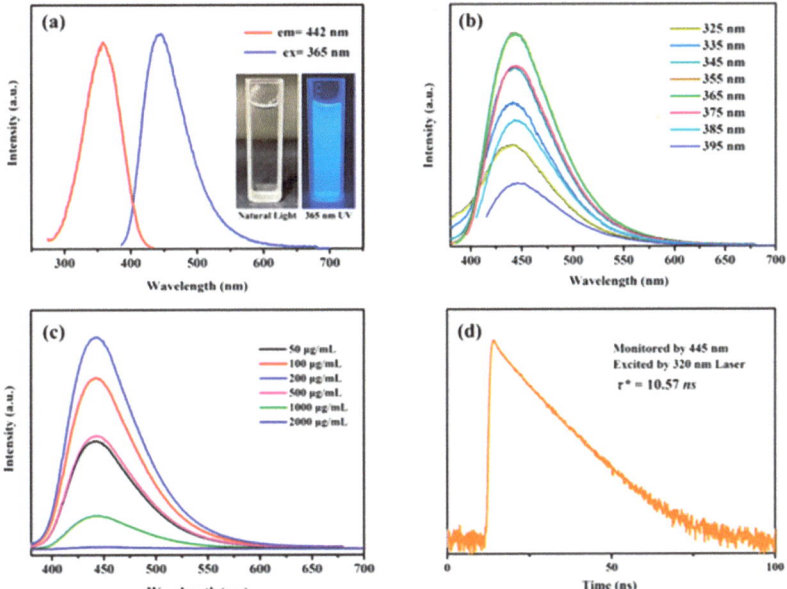

Figure 5. (a) Excitation and emission spectra of 100 μg mL^{-1} cys-CD aqueous solution. (b) Emission spectra of 100 μg mL^{-1} cys-CD aqueous solution at different excitation wavelengths. (c) cys-CD fluorescence spectra of aqueous solutions with different concentrations of cys-CD. (d) Fluorescence lifetime decay of cys-CD.

The pH dependence in the UV-vis responses is the protonation/deprotonation of the N-centers present in the CD honeycomb matrix [21]. The cys-CD emission wavelength of 442 nm exhibited no dependence on excitation within the range of 325–375 nm (Figure 5b). This indicates uniformity in both the types and sizes of the functional groups attached to the cys-CD surface. Such uniformity in emission sites is advantageous for future applications involving CD fluorescence.

We measured the fluorescence of the cys-CD aqueous solutions at various concentrations under excitation at a wavelength of 365 nm (Figure 5c). The intensity of the cys-CD fluorescence was directly proportional to its concentration within the range of 50–200 μg mL^{-1}. However, as the concentration increased beyond 200 μg mL^{-1} up to 2000 μg mL^{-1}, the fluorescence intensity decreased gradually. This phenomenon can be attributed to self-quenching, where at higher concentrations, molecules tend to accumulate, leading to a reduction in fluorescence intensity.

Thus, we determined that the optimal blue-fluorescence intensity in aqueous solution under neutral conditions was achieved with excitation wavelengths ranging from 355 nm to 365 nm, at a cys-CD concentration of 100–200 μg mL^{-1} (Figure 5c). Additionally, the fluorescence lifetime of cys-CD was measured to be 10.57 ns under laser excitation at a wavelength of 320 nm, with monitoring at an emission wavelength of 445 nm. From the attenuation curve of the fluorescence intensity in Figure 5d, when the fluorescent substance is excited, most of the excited state molecules return to the ground state rapidly. But, some molecules return to the ground state in several times the fluorescence lifetime delay [22]. Besides, because of the intermolecular interaction or other situations, the excited stated energy is consumed, which causes the fluorescence quenching. It is also called aggregation-induced fluorescence quenching.

2.3. Stability of cys-CD Fluorescence

The fluorescence intensity of a 100 μg mL^{-1} solution of cys-CD in water remained relatively stable across a range of Na$^+$ and Cl$^-$ concentrations from 0 to 400 mM (Figure 6a). This indicates that the fluorescence of cys-CD is highly resistant to salt interference. We then investigated the pH stability of the fluorescence in a standard solution of 100 μg mL^{-1} cys-CD over the pH range of 2–12 (Figure 6b). The highest blue-fluorescence intensity of cys-CD was observed at pH 7. Even under slightly acidic or basic conditions, cys-CD maintained its blue-fluorescence properties well. However, at pH < 3 (strongly acidic conditions), the fluorescence intensity of cys-CD decreased significantly, accompanied by a red shift in the emission wavelength. Similarly, at pH > 11 (strongly basic conditions), the fluorescence intensity decreased significantly, although there was no red shift in the emission wavelength.

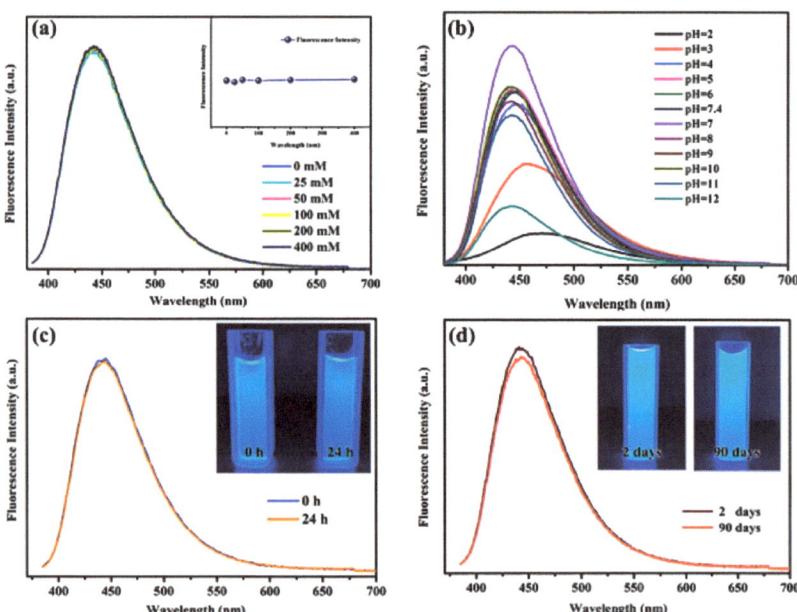

Figure 6. (a) Fluorescence spectra of cys-CD in 0–400 mM NaCl solutions. (b) The fluorescence spectra of cys-CD in the pH range of 2–12. (c) The cys-CD fluorescence spectra before and after exposure to a 365 nm UV lamp for 24 h. (d) The cys-CD fluorescence spectra before and after 90 days of storage.

One possible explanation is that the functional groups on the surface of cys-CDs are easily hydrolyzed in strong acidic or basic environments, leading to a decrease in overall fluorescence. Additionally, at low pH levels, the protonation of certain groups may modulate the absorption maximum.

Following 24 h of UV irradiation at 365 nm with a power of 12 W, the fluorescence intensity of a 100 μg mL^{-1} cys-CD aqueous solution showed no significant reduction (refer to Figure 6c and the inset). This suggests that cys-CDs demonstrate robust stability and resistance to photobleaching. To further assess stability, the fluorescence of a 100 μg mL^{-1} cys-CD aqueous solution was examined after 90 days of storage at 4 °C. The comparison of the fluorescence spectra between fresh and stored solutions, measured under identical conditions (following treatment in a water bath at 25 °C for 15 min), revealed no significant changes (Figure 6d). This demonstrates that cys-CDs maintain stable fluorescence over extended periods, indicating their long-term stability.

2.4. Baicalein Detection

Following the optimization of detection conditions using cys-CD fluorescence on baicalein, concentrations ranging from 0 μM to 30 μM were measured under optimal conditions (cys-CD concentration: 5 μg mL^{-1}; PBS buffer pH 7.4; reaction time: 10 min). The test results, depicted in Figure 7, illustrated a decrease in the cys-CD fluorescence intensity with an increasing baicalein concentration. The fluorescence intensity corresponding to each baicalin concentration served as the focus of the research.

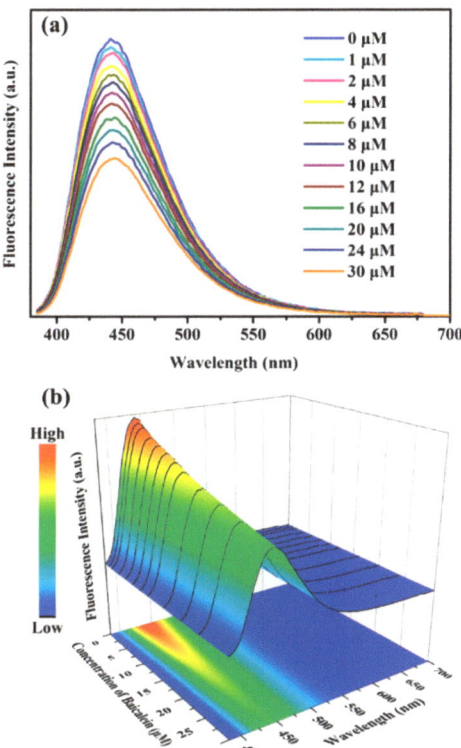

Figure 7. Variation in the fluorescence of cys-CD (5 μg mL^{-1}) with different concentrations of baicalein in the system. (**a**) The 2D spectra. (**b**) The 3D spectra.

In Figure 8, the Lineweaver–Burk curve is presented, wherein the integral area of the spectral curve was calculated. The fluorescence intensity differences between the blank and dosing groups, represented as ($F_0 - F$), were obtained. Their reciprocal ($1/(F_0 - F)$)

demonstrated a strong linear relationship with the reciprocal ($1/C_q$) of the baicalin concentration, aligning with the Lineweaver–Burk equation: $\frac{1}{F_0-F} = \frac{1}{F_0} + \frac{K_{lb}}{F_0 C_q}$, with a linear correlation coefficient (R^2) of 0.998.

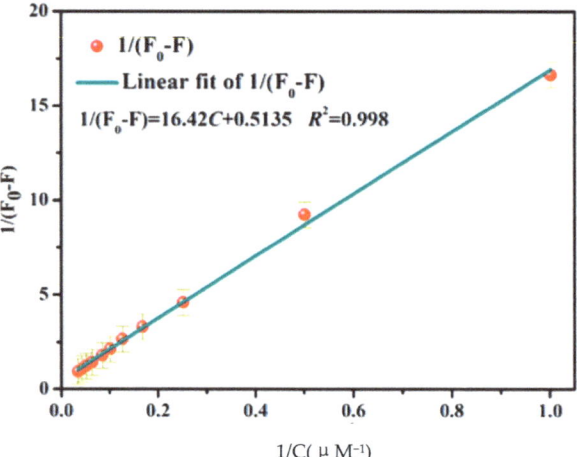

Figure 8. Lineweaver–Burk curve of baicalein detection by cys-CD.

The detection limit (*LOD*) for baicalein was calculated to be 33 nM using Equation (1), where *N*, *Q*, and *I* represent the noise level, the injection volume, and the signal value, respectively. Baicalein at a concentration of 10 µM was measured 15 times, resulting in a relative standard deviation (*RSD*) of 6.82% for this detection method.

$$LOD = 3N \times \frac{Q}{I} \qquad (1)$$

2.5. Mechanism Verification of Baicalein Detection Using cys-CDs

Fluorescence quenching encompasses dynamic and static quenching processes, each with distinct mechanisms [23,24]. Static quenching occurs when a quenching moiety forms a complex with a fluorescent molecule in its ground state, preventing it from transitioning to the excited (fluorescent) state [25,26]. In static quenching, the rate of the decay of the excited state due to fluorescence remains unchanged, meaning that the fluorescence lifetime remains constant [27]. The Lineweaver–Burk equation (Equation (2)) describes static quenching.

$$\frac{F_0}{F} = 1 + CK_{SV} \qquad (2)$$

Dynamic quenching happens when fluorescent moieties in an excited state collide with other molecules that can interfere with fluorescence, either by facilitating alternative relaxation pathways like energy or charge transfer transitions, causing the excited state to return to its ground state [28–30]. Consequently, this alters the duration of the excited state. The Stern–Volmer equation (Equation (3)) describes dynamic quenching.

$$\frac{1}{F_0 - F} = \frac{1}{F_0} + \frac{K_{lb}}{F_0 C_q} \qquad (3)$$

2.6. Fluorescence Inner-Filter Effect

When a system contains both a fluorophore and a non-fluorescent absorber, the absorption spectrum of the absorbing material may coincide with the excitation and emission spectra of the fluorophore [31,32]. This overlap reduces the number of excitation photons reaching the detection volume of the apparatus, leading to a reduced fluorescence intensity. This phenomenon is known as the fluorescence inner-filter effect. For baicalein and cys-CD, the UV-Vis absorption spectrum of baicalein overlaps with the excitation and emission spectra of cys-CD (Figure 9), potentially causing the fluorescence quenching of cys-CD by baicalein.

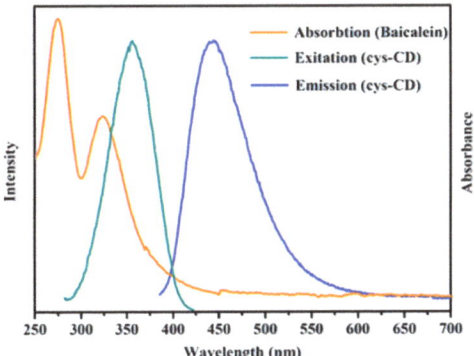

Figure 9. UV-Vis absorption of baicalein and excitation–emission spectra of cys-CD.

2.7. Static Quenching Effect

From Figure 10, there are three peaks (ca 250 nm p-p* transitions, ca 320 nm, and ca 370 nm n-p* transitions) in the UV-vis spectrum of CDs. The transition of the C=O and C=N groups are both n-p* transitions [33]. Changes in the UV-Vis absorption spectrum of cys-CD were observed before and after the addition of baicalein, as depicted in Figure 10a. Following the addition of baicalein, it was evident that the UV absorption spectrum diverged notably from that of the neat cys-CD solution. Furthermore, the absorption peak broadened, indicating the emergence of new complexes within the system, as illustrated in Figure 10b.

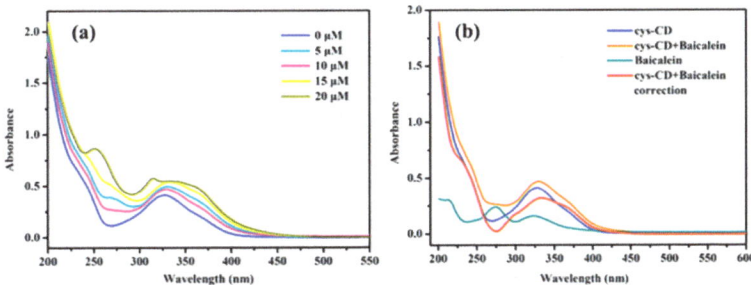

Figure 10. (a) Effect of different concentrations of baicalein on cys-CD's UV absorption. (b) Effect of 10 μM baicalein on cys-CD's UV absorption.

Subsequently, we investigated the impact of the various concentrations of baicalein on the fluorescence lifetime of cys-CD, as shown in Figure 11. In the presence of baicalein within the concentration range of 0–30 μM, the fluorescence lifetime of cys-CD remained virtually constant at 10.53 ± 0.32 ns. This suggests that there was minimal dynamic quenching occurring.

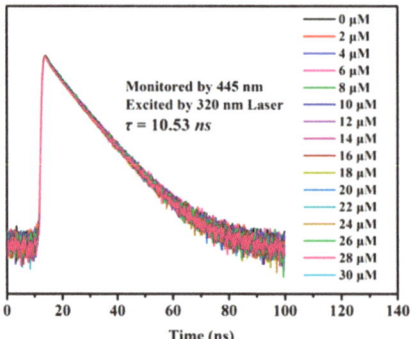

Figure 11. Fluorescence-decay of cys-CD with different concentrations of baicalein.

In summary, the experimental findings suggest that the fluorescence quenching of cys-CD by baicalein is attributable to both static quenching and an inner-filter effect.

2.8. Baicalein Detection in Biological Samples

In this experiment, we employed the standard sampling method. We spiked human urine and blood samples with baicalein at known concentrations and compared them to the measured concentrations. This process verified the feasibility of detecting baicalein using cys-CD in real-world samples [34]. The recovery values of this detection method in human urine and blood samples ranged from 97.10% to 98.19% and from 101.93% to 103.49%, respectively (Table 1). These results demonstrate the reliability of our baicalein detection method.

Table 1. Recovery of baicalein by cys-CD in human urine and human serum samples.

Sample	Group	Concentration of Baicalein (mM)		Recovery (%)	RSD (%) n = 5
		Spiked	Found		
Serum	1	8	7.85	98.19	3.81
	2	10	9.71	97.10	4.09
	3	12	11.74	97.87	2.44
Urine	1	8	8.28	103.49	3.52
	2	10	10.25	102.5	1.91
	3	12	12.23	101.93	2.77

3. Materials and Methods

3.1. Baicalein Detection Methods

Baicalein, the primary active ingredient found in *Scutellaria baicalensis*, serves as a crucial indicator for evaluating the quality of this herb. Currently available methods for measuring baicalein content primarily include spectrometry, chromatography, and biological methods. Below, we outline the advantages and disadvantages of each approach:

3.1.1. UV-Vis Spectrophotometry

UV-Vis spectrophotometry is a commonly employed technique for quantifying baicalein content. It relies on the absorption characteristics of baicalein at specific wavelengths, typically set at 260 nm. This method is straightforward and requires minimal sample pre-treatment. However, in real-world applications, accuracy can be compromised by interference from the sample matrix. Consequently, prior to measurement, samples must undergo pre-treatment to mitigate this interference.

3.1.2. Mass Spectrometry

Mass spectrometry (MS) is a method used to determine the structure and mass–charge ratio of molecular fragments. It finds extensive application in detecting the content of baicalein. Liquid chromatography–mass spectrometry, including LC-MS/MS, and gas chromatography–mass spectrometry (GC-MS) are commonly employed for this purpose. Mass spectrometry allows for the direct detection of molecular information pertaining to baicalein, even in complex samples, without being affected by interference from other substances. Consequently, the measurement results are highly accurate and reliable. However, this method requires sophisticated facilities and precise operational procedures. Additionally, it is relatively costly, making it less commonly used in general experiments.

3.1.3. High-Performance Liquid Chromatography

High-performance liquid chromatography (HPLC) stands out as one kind of the most commonly used and highly accurate methods for detecting baicalein content. This method primarily relies on a reverse-chromatographic column for separation, utilizing an ethanol-water mixture as the mobile phase. The determination process involves both gradient elution and isothermal elution. Typically, a UV detector set at 254 nm is employed for testing, ensuring accurate and reliable results across a broad range of applications.

However, this method notably demands sophisticated facilities and precise operational procedures. Throughout the detection process, strict control over various factors is essential to maintain the accuracy and reliability of the test results. The baicalein content detection method described in this manuscript utilizes HPLC.

3.2. Reagents and Equipment

The reagents used in these experiments are listed in Table 2.

Table 2. Reagents used in this study.

Reagents	Purity	Manufacturer
L-cysteine	Analytical pure	Aladdin Reagent Co., Ltd. (Shanghai, China)
Sodium citrate dihydrate	Analytical pure	GHTECH Technology Co., Ltd. (Guangzhou, China)
Baicalein standard	99.8%	Aladdin Reagent Co., Ltd. (Shanghai, China)
Sodium chloride	Analytical pure	Damao Chemical Reagent Factory (Tianjin, China)
Acetone	Analytical pure	Damao Chemical Reagent Factory (Tianjin, China)
EDTA	Analytical pure	Damao Chemical Reagent Factory (Tianjin, China)
Absolute alcohol	Analytical pure	GHTECH Technology Co., Ltd. (Guangzhou, China)
CTAB	Analytical pure	Aladdin Reagent Co., Ltd. (Shanghai, China)
Phosphate buffer (pH 7.4)	Analytical pure	Yuanye Bio-Technology Co., Ltd. (Shanghai, China)
DEME culture medium	Analytical pure	Tianjun Bio-Technology Co., Ltd. (Guangzhou, China)
4% PFA stationary liquid	Analytical pure	Bairui Bio-Technology Co., Ltd. (Shanghai, China)
Anti-fluorescence quenching mounting medium	Analytical pure	Cida Bio-Technology Co., Ltd. (Guangzhou, China)

The used instruments are outlined in Table 3.

Table 3. Instruments used in this study.

Instruments	Model	Manufacturer
Steady/Transient spectrometer	FLS980	Edinburgh Instrument Company (Edinburgh, Britain)
FTIR	6700 FT-IR	Thermo Fisher Scientific Technology Company (Waltham, MA, USA)
UV-Visible-near infrared light spectrophotometer	UV-3600	Shimadzu Corporation (Kyoto, Japan)
TEM	FEI TECNAI G2 F20	FEI Company (Waltham, MA, USA)
Multifunctional XPS	AXIS ULTRA DLD	Shimadzu Corporation (Kyoto, Japan)

Table 3. Cont.

Instruments	Model	Manufacturer
Liquid nitrogen cryostat	OptistatDN-V2	Oxford Instruments Technology Co., Ltd. (Shanghai, China)
Thermostatic magnetic stirrer	85-2	Aohua Instruments Co., Ltd. (Changzhou, China)
High-speed centrifuge	TG16-WS	Xiangyi Laboratory Instrument Development Co., Ltd. (Changsha, China)
Thermal-Storage heating magnetic stirrer	DF-101S	Yuhua Instruments Co., Ltd. (Gongyi, China)
Analytical balance	AX124 ZH/E	Ohaus Corporation (Newark, NJ, USA)
UV analyzer	ZF-20D	Yuhua Instruments Co., Ltd. (Gongyi, China)
Electric blast drying oven	DHG-9145A	Aohua Instruments Co., Ltd. (Changzhou, China)
Laser confocal microscope	LSM 880 with Airyscan	Xiangyi Laboratory Instrument Development Co., Ltd. (Changsha, China)
Freeze dryer	Freezone6L	Shimadzu Corporation (Kyoto, Japan)

3.3. Synthesis and Purification of Blue-Fluorescent S, N-co-Doped CDs (cys-CD)

L-cysteine served as a source of carbon, nitrogen, and sulfur, while sodium citrate acted as a secondary carbon source. Ethylenediamine (EDA) was utilized to control the shape [35]. The specific synthesis steps for cys-CD are outlined below: Initially, 0.2 g of L-cysteine and 0.2 g of sodium citrate were weighed and dissolved in 10 mL of ultrapure water. Subsequently, 500 µL of anhydrous EDA was added to the mixture, which was then sonicated for 5 min. The resulting colorless and transparent mixture was transferred to a 40 mL polytetrafluoroethylene (PTFE) reactor lining within a stainless steel hydrothermal reactor. This assembly was heated in an electric blast drying oven at 200 °C for 4 h. Once the reaction was complete, the mixture was allowed to cool and then filtered using a 0.22 µm Millipore filter. The filtrate was then poured into a dialysis bag with a 1000 Da molecular weight cut-off (MWCO) and dialyzed at 4 °C for 48 h to obtain purified cys-CD. To further investigate the relationship between CD concentration and fluorescence properties, the purified cys-CD underwent freeze-drying for 24 h. Finally, the resulting freeze-dried cys-CD powder was dissolved in ultrapure water, adjusted to a concentration of 20 mg mL^{-1}, and stored at 4 °C for subsequent use.

3.4. Fluorescence Properties of cys-CD

To ensure a consistent comparison of the experimental results, the fluorescence properties of cys-CD solutions were examined using an FLS980 fluorometer under standardized conditions. A solution of cys-CD (3 mL) was added to a cuvette. The excitation wavelength was 365 nm. The slit width was 1.2 nm. The residence time was set at 0.1 s to record the emission spectrum over the wavelength range of 380–700 nm.

Aqueous solutions of cys-CDs were prepared at 100 and 2000 µg mL^{-1}. For spectral testing, 3 mL of the 100 µg mL^{-1} concentration was used. The excitation wavelength range tested spanned from 325 to 395 nm.

NaCl solution (0–400 mM) was used to prepare the cys-CD solution (100 µg mL^{-1}) used for the fluorescence spectra scanning. NaOH (2 M) and HCl (1 M) solutions were used to adjust the pH of the PBS buffer between 2 and 12. PBS solutions at different pH levels were used to prepare the cys-CD solution (100 µg mL^{-1}); 3 mL of this solution was then used for fluorescence spectra scanning.

The same solution sample was irradiated by exposure to a UV lamp (365 nm, 12 W) for 24 h before being used for the fluorescence spectral testing. Simultaneously, a cys-CD solution (2000 µg mL^{-1}) was stored at 4 °C for 90 days and subsequently used for the fluorescence spectral testing.

3.5. cys-CD Fluorescence Quantum Yield

The fluorescence quantum yield (FQY) of a material refers to the ratio of the absorbed photons to the emitted photons via fluorescence, typically measured at a specific excitation wavelength. This parameter is crucial for evaluating the photoelectric conversion capability of fluorescent materials, with higher values indicating superior optical properties. In this test, a comparative measurement method was employed using quinine sulfate (FQY = 0.56) as the standard reference. The emission and UV-Vis spectra of both quinine sulfate and cys-CD were recorded, followed by the calculation of the quantum yield of cys-CD using Equation (4),

$$FQY_x = FQY_{st} \times \frac{I_x}{I_{st}} \times \frac{\eta_x^2}{\eta_{xt}^2} \times \frac{A_{st}}{A_x} \quad (4)$$

where x represents the sample, st is the standard (e.g., quinine sulfate), I is the integral of the fluorescence emission spectrum, η is the solvent refractive index, and A is absorbance.

3.6. Fluorescence Lifetime

We utilized an FLS980 transient steady-state fluorometer for our experiment, employing a 320 nm laser for excitation. The monitoring wavelength was set at 442 nm to measure the fluorescence lifetime of cys-CD. Equation (5) was used to calculate the fluorescence lifetime.

$$I(t) = I_0 + A_1 exp\left(-\frac{t}{\tau_1}\right) + A_2 exp\left(-\frac{t}{\tau_2}\right) \quad (5)$$

In Equation (5), $I(t)$ is the fluorescence intensity of the sample at 442 nm. I_0 is the initial fluorescence intensity at $t = 0$. A is the fluorescence lifetime constant. τ is the fluorescence lifetime decaying exponential. The average fluorescence lifetime (τ^*) was then calculated using Equation (6).

$$\tau^* = \frac{\int_0^\infty tI(t)dt}{\int_0^\infty I(t)dt} \quad (6)$$

3.7. Optimization of Baicalein Measurement Conditions

3.7.1. Effect of cys-CD Concentration

To examine how the concentration of cys-CD affects the sensitivity of the baicalein response, we prepared aqueous solutions of cys-CD at the concentrations of 30, 40, 50, 60, and 70 μg mL^{-1}, each totaling 500 μL. To these solutions, 2 mL of absolute ethanol was added. Subsequently, 2 mL of PBS buffer at pH 7.4 was thoroughly mixed. This resulting buffer was then combined with a baicalein ethanol solution (1 mM) in various volumes, with the total volume kept constant at 5 mL by adding absolute ethanol as necessary. The concentration of the baicalein in the test samples ranged from 0 to 20 μM. After thorough mixing, the solution was allowed to equilibrate for 10 min, following which its fluorescence spectrum was measured.

3.7.2. In Vitro pH Effects

To investigate how pH affects the fluorescence of cys-CD in detecting baicalein, we prepared multiple 4 mL solutions by combining 2 mL of absolute ethanol with 2 mL of PBS buffer at pH 7.4. We adjusted the pH of each solution to 4, 5, 6, 7, 8, 9, 10, and 11 by adding small amounts of 2 M NaOH and 1 M HCl solutions. Then, 4 mL of each solution was mixed with 500 μL of cys-CD solution (50 μg mL^{-1}) and varying volumes of baicalein ethanol solution (1 mM). To maintain a constant volume of 5 mL, absolute ethanol was added accordingly. After thorough mixing, the samples were allowed to equilibrate for 10 min before measuring the fluorescence spectrum.

3.7.3. Effect of Reaction Time

To investigate how reaction time influences baicalein detection, we set the cys-CD concentration to 5 μg mL^{-1} and the baicalein concentration to 10 μM. The experiment was conducted at pH 7.4. Following the addition of the baicalein, we measured the fluorescence spectrum at various time intervals ranging from 0 to 60 min.

3.7.4. Effect of Ionic Strength

To investigate how ionic strength influences the detection of baicalein, we prepared solutions containing 5 μg mL^{-1} cys-CD and 10 μM baicalein at pH 7.4. We then added NaCl to achieve the concentrations of 0, 25, 50, 100, 200, and 400 mM, respectively. After shaking and allowing the solutions to equilibrate for 10 min, we measured the fluorescence spectrum.

3.7.5. Baicalein Detection

First, a 500 μL aliquot of a cys-CD aqueous solution at 50 μg mL^{-1} concentration was combined with 2 mL of absolute ethanol. Subsequently, 2 mL of PBS buffer (pH 7.4) was thoroughly mixed and added to varying volumes of a baicalein (1 mg mL^{-1}) ethanol solution at 1 mg mL^{-1} concentration. Absolute ethanol was added to maintain a constant total volume of 5 mL. The resulting baicalein test solutions were prepared across concentrations ranging from 0 to 30 μM. After shaking and allowing for a 10 min equilibration period, the fluorescence spectra of each solution were recorded.

3.8. Mechanism of Baicalein Detection

3.8.1. Effect of Baicalein on cys-CD UV-Vis Absorbance Spectrum

A solution containing cys-CD (50 μg mL^{-1}) was mixed with the baicalein solutions of different concentrations (0, 5, 10, 15, and 20 μM) for 10 min. Subsequently, the UV-Vis absorption spectrum was recorded over the wavelength range of 200–700 nm.

3.8.2. Effect of Baicalein on cys-CD Fluorescence Lifetime

Aqueous solutions containing cys-CD at a concentration of 50 μg mL^{-1} were exposed to the baicalein solutions of concentrations ranging from 0 to 30 μM for 10 min. Subsequently, the fluorescence lifetime was determined under identical conditions outlined in Section 3.8.1.

3.8.3. Detection of Baicalein in Urine and Blood Using cys-CD

In this experiment, we used urine and blood samples from healthy volunteers to investigate the relationship between the concentration of the added baicalein and the resulting concentration detected.

For urine samples, we began by centrifuging them for 15 min at 8000 rpm, then preserving the supernatant. This supernatant was then diluted 500 times with ultrapure water and stored in the refrigerator until needed.

Blood plasma samples were mixed with acetonitrile in a 1:1 ratio, followed by a 5 min equilibration period to remove proteins. After centrifugation for 10 min at 10,000 rpm, the supernatant was extracted and filtered using a polyethersulfone (PES) ultrafiltration membrane with a pore diameter of 0.45 μm. The resulting filtrate was diluted 100 times in PBS buffer (pH 7.4) and was refrigerated until use.

Using the standard addition method, we added baicalein standard solutions to both urine and blood samples. Finally, we measured the concentration of each mixture.

4. Conclusions

L-cysteine and sodium citrate were utilized as raw materials in the hydrothermal synthesis process to produce blue-fluorescent S, N-co-doped carbon dots (cys-CDs). These cys-CDs exhibited a remarkably high fluorescence quantum yield of 0.66. The characterization of the material was performed using various techniques including TEM, FTIR, XPS, UV-Vis, and PL/PLE, to analyze particle size distribution, structural composition, and fluorescence properties.

A rapid, simple, and sensitive method for detecting baicalein was developed based on the static-quenching effect of baicalein on cys-CD fluorescence. Baicalein concentrations within the concentration range of 0–30 μM exhibited a linear relationship with the extent of cys-CD fluorescence quenching, with a detection limit of 33 nM. This approach was successfully applied to detect baicalein in human urine and blood samples, yielding satisfactory recovery concentrations.

In summary, we introduced an environmentally sustainable and efficient method for synthesizing and purifying cys-CDs. Additionally, we conducted a preliminary investigation into the potential application of cys-CDs for drug testing. These findings suggest promising applications of cys-CDs in pharmaceutical, biosecurity, and medical imaging contexts.

Author Contributions: Conceptualization, Y.C. and G.Y.; methodology, G.Y.; software, Y.H.; validation, G.Y.; formal analysis, G.Y.; investigation, Y.C.; resources, Y.H.; data curation, G.Y.; writing—original draft preparation, G.Y.; writing—review and editing, G.Y.; visualization, Y.C.; supervision, Y.H.; project administration, Y.C.; funding acquisition, Y.H. All authors have read and agreed to the published version of the manuscript.

Funding: This research was aided by the key area campaign of regular universities, No. 2021ZDZX1058. Guangdong Basic and Applied Basic Research Foundation, No. 2023A1515240063. Guangdong universities featured innovation project, No. 2022KTSCX194.

Data Availability Statement: Data are contained within the article.

Conflicts of Interest: The authors declare no conflicts of interest.

References

1. Zhang, Y.; Yao, H.; Xu, Y.; Xia, Z. Synergistic Weak/Strong Coupling Luminescence in Eu-Metal-Organic Framework/Zn_2GeO_4:Mn^{2+} Nanocomposites for Ratiometric Luminescence Thermometer. *Dyes Pigments* **2018**, *157*, 321–327. [CrossRef]
2. Zhou, Y.; Zhang, D.; Zeng, J.; Gan, N.; Cuan, J. A Luminescent Lanthanide-Free MOF Nanohybrid for Highly Sensitive Ratiometric Temperature Sensing in Physiological Range. *Talanta* **2018**, *181*, 410–415. [CrossRef] [PubMed]
3. Gharari, Z.; Bagheri, K.; Danafar, H.; Sharafi, A. Simultaneous Determination of Baicalein, Chrysin and Wogonin in Four Iranian *Scutellaria* species by High Performance Liquid Chromatography. *J. Appl. Res. Med. Aromat. Plants* **2020**, *16*, 100232. [CrossRef]
4. Zhang, X.; Zhu, Z.; Guo, Z.; Sun, Z.; Chen, Y. A Ratiometric Optical Thermometer with High Sensitivity and Superior Signal Discriminability Based on $Na_3Sc_2P_3O_{12}$: Eu^{2+}, Mn^{2+} Thermochromic Phosphor. *Chem. Eng. J.* **2019**, *356*, 413–422. [CrossRef]
5. Zhang, X.; Zhu, Z.; Guo, Z.; Sun, Z.; Yang, Z.; Zhang, T.; Zhang, J.; Wu, Z.C.; Wang, Z. Dopant Preferential Site Occupation and High Efficiency White Emission in $K_2BaCa(PO_4)_2$:Eu^{2+},Mn^{2+} Phosphors for High Quality White LED Applications. *Inorg. Chem. Front.* **2019**, *6*, 1289–1298. [CrossRef]
6. Bai, Y.M.; Mao, J.; Li, D.X.; Luo, X.J.; Chen, J.; Tay, F.R.; Niu, L.N. Bimodal Antibacterial System Based on Quaternary Ammonium Silane-Coupled Core-Shell Hollow Mesoporous Silica. *Acta Biomater.* **2019**, *85*, 229–240. [CrossRef] [PubMed]
7. An, G.S.; Shin, J.R.; Hur, J.U.; Oh, A.H.; Kim, B.; Jung, Y.; Choi, S. Fabrication of Core-Shell Structured Fe_3O_4@Au Nanoparticle via Self-Assembly Method Based on Positively Charged Surface Silylation/Polymerization. *J. Alloys Compd.* **2019**, *798*, 360–366. [CrossRef]
8. Bardi, B.; Tosi, I.; Faroldi, F.; Baldini, L.; Sansone, F.; Sissa, C.; Terenziani, F. A Calixarene-Based Fluorescent Ratiometric Temperature Probe. *Chem. Commun.* **2019**, *55*, 8098–8101. [CrossRef] [PubMed]
9. Wu, Z.; Tan, B.; Velasco, E.; Wang, H.; Shen, N.N.; Guo, Y.J.; Zhang, X.; Zhu, K.; Zhang, G.Y.; Liu, Y.Y.; et al. Fluorescent in Based MOFs Showing "Turn On" Luminescence towards Thiols and Acting as a Ratiometric Fluorescence Thermometer. *J. Mater. Chem. C* **2019**, *7*, 3049–3055. [CrossRef]
10. Macairan, J.R.; Jaunky, D.B.; Piekny, A.; Naccache, R. Intracellular Ratiometric Temperature Sensing Using Fluorescent Carbon Dots. *Nanoscale Adv.* **2019**, *1*, 105–113. [CrossRef]
11. Liu, X.; Liu, J.; Zhou, H.; Yan, M.; Liu, C.; Guo, X.; Xie, J.; Li, S.; Yang, G. Ratiometric Dual Fluorescence Tridurylboron Thermometers with Tunable Measurement Ranges and Colors. *Talanta* **2020**, *210*, 120630. [CrossRef]
12. Yin, P.; Niu, Q.; Yang, Q.; Lan, L.; Li, T. A New "Naked-Eye" Colorimetric and Ratiometric Fluorescent Sensor for Imaging Hg^{2+} in Living Cells. *Tetrahedron* **2019**, *75*, 130687. [CrossRef]
13. Wu, X.Y.; Zhao, Q.; Zhang, D.X.; Liang, Y.C.; Zhang, K.K.; Liu, Q.; Dong, L.; Shan, C.X. A Self-Calibrated Luminescent Thermometer Based on Nanodiamond-Eu/Tb Hybrid Materials. *Dalton Trans.* **2019**, *48*, 7910–7917. [CrossRef] [PubMed]
14. Gao, H.; Kam, C.; Chou, T.Y.; Wu, M.Y.; Zhao, X.; Chen, S. A Simple yet Effective AIE-Based Fluorescent Nano-thermometer for Temperature Mapping in Living Cells Using Fluorescence Lifetime Imaging Microscopy. *Nanoscale Horiz.* **2020**, *5*, 488–494. [CrossRef]
15. He, S.; Qi, S.; Sun, Z.; Zhu, G.; Zhang, K.; Chen, W. Si, N-codoped Carbon Dots: Preparation and Application in Iron Overload Diagnosis. *J. Mater. Sci.* **2019**, *54*, 4297–4305. [CrossRef]
16. Xia, W.; Zhang, W. Characterization of surface properties of Inner Mongolia coal using FTIR and XPS. Energy sources part a-recovery utilization and environmental effects. *Energy Sources Part A Recover. Util. Environ. Eff.* **2017**, *39*, 1190–1194.

17. Zhu, S.J.; Meng, Q.N.; Wang, L.; Zhang, J.; Song, Y.; Jin, H.; Zhang, K.; Sun, H.; Wang, H.; Yang, B. Highly photoluminescent carbon dots for multicolor pattrening, sensors, and bioimaging. *Angew. Commun.* **2013**, *52*, 3953–3957. [CrossRef] [PubMed]
18. Ye, Y.; Yang, D.; Chen, H.; Guo, S.; Yang, Q.; Chen, L.; Zhao, H.; Wang, L. A high-efficiency corrosion inhibitor of N-doped citric acid-based carbon dots for mild steel in hydrochloric acid environment. *J. Hazard. Mater.* **2019**, *2020*, 121019. [CrossRef]
19. Wang, C.; Wang, Y.; Shi, H.; Yan, Y.; Liu, E.; Hu, X.; Fan, J. A strong blue fluorescent nanoprobe for highly sensitive and selective detection of mercury(II) based on sulfur doped carbon quantum dots. *Mater. Chem. Phys.* **2019**, *232*, 145–151. [CrossRef]
20. Guo, D.; Dong, Z.; Luo, C.; Zan, W.; Yan, S.; Yao, X. A rhodamine B-based "turn-on" fluorescent sensor for detecting Cu^{2+} and sulfur anions in aqueous media. *RSC Adv.* **2014**, *4*, 5718–5725. [CrossRef]
21. Vercelli, B.; Donnini, R.; Ghezzi, F.; Sansonetti, A.; Giovanella, U.; La Ferla, B. Nitrogn-doped carbon quantum dots obtained hydrothermally from citric acid and urea: The role of the specific nitrogen centers in their electrochemical and optical responses. *Electrochim. Acta* **2021**, *387*, 138557. [CrossRef]
22. Inada, N.; Fukuda, N.; Hayashi, T.; Uchiyama, S. Temperature Imaging Using a Cationic Linear Fluorescent Polymeric Thermometer and Fluorescence Lifetime Imaging Microscopy. *Nat. Protoc.* **2019**, *14*, 1293–1321. [CrossRef]
23. Nguyen, V.; Si, J.H.; Yan, L.H.; Hou, X. Electron-hole recombination dynamics in carbon nanodots. *Carbon* **2015**, *95*, 659–663. [CrossRef]
24. Wang, W.; Peng, J.; Li, F.; Su, B.; Chen, X.; Chen, X. Phosphorus and Chlorine Co-Doped Carbon Dots with Strong Photoluminescence as a Fluorescent Probe for Ferric Ions. *Mikrochim. Acta* **2018**, *186*, 32. [CrossRef] [PubMed]
25. Wang, S.; Cao, J.; Lu, C. A Naphthalimide-Based Thermometer: Heat-Induced Fluorescence "Turn-On" Sensing in a Wide Temperature Range in Ambient Atmosphere. *New J. Chem.* **2020**, *44*, 4547–4553. [CrossRef]
26. Yang, F.; He, X.; Wang, C.; Cao, Y.; Li, Y.; Yan, L.; Liu, M.; Lv, M.; Yang, Y.; Zhao, X.; et al. Controllable and Eco-Friendly Synthesis of P-Riched Carbon Quantum Dots and Its Application for Copper (II) Ion Sensing. *Appl. Surf. Sci.* **2018**, *448*, 589–598. [CrossRef]
27. Yang, W.; Zhang, H.; Lai, J.; Peng, X.; Hu, Y.; Gu, W.; Ye, L. Carbon Dots with Red-shifted Photoluminescence by Fluorine Doping for Optical Bio-Imaging. *Carbon* **2018**, *128*, 78–85. [CrossRef]
28. He, X.; Luo, Q.; Zhang, J.; Chen, P.; Wang, H.J.; Luo, K.; Yu, X.Q. Gadolinium-Doped Carbon Dots as Nano-Theranostic Agents for MR/FL Diagnosis and Gene Delivery. *Nanoscale* **2019**, *11*, 12973–12982. [CrossRef] [PubMed]
29. Das, P.; Maity, P.P.; Ganguly, S.; Ghosh, S.; Baral, J.; Bose, M.; Choudhary, S.; Gangopadhyay, S.; Dhara, S.; Das, A.K.; et al. Biocompatible Carbon Dots Derived from Kappa-Carrageenan and Phenyl Boronic Acid for Dual Modality Sensing Platform of Sugar and Its Anti-diabetic Drug Release Behavior. *Int. J. Biol. Macromol.* **2019**, *132*, 316–329. [CrossRef]
30. Zhao, J.; Zheng, Y.; Pang, Y.; Chen, J.; Zhang, Z.; Xi, F.; Chen, P. Graphene Quantum Dots as Full-Color and Stimulus Responsive Fluorescence Ink for Information Encryption. *J. Colloid Interface Sci.* **2020**, *579*, 307–314. [CrossRef]
31. Wang, H.; Wei, J.; Zhang, C.; Zhang, Y.; Zhang, Y.; Li, L.; Yu, C.; Zhang, P.; Chen, J. Red Carbon Dots as Label-Free Two-Photon Fluorescent Nanoprobes for Imaging of Formaldehyde in Living Cells and Zebrafishes. *Chin. Chem. Lett.* **2020**, *31*, 759–763. [CrossRef]
32. Yan, F.; Bai, Z.; Zu, F.; Zhang, Y.; Sun, X.; Ma, T.; Chen, L. Yellow-Emissive Carbon Dots with a Large Stokes Shift Are Viable Fluorescent Probes for Detection and Cellular Imaging of Silver Ions and Glutathione. *Mikrochim. Acta* **2019**, *186*, 113. [CrossRef] [PubMed]
33. Meng, Y.; Yang, M.; Liu, X.; Yu, W.; Yang, B. Zn^{2+}-Doped Carbon Dots, a Good Biocompatibility Nanomaterial Applied for Bio-imaging and Inducing Osteoblastic Differentiation In Vitro. *ACS Nano* **2019**, *14*, 1950029. [CrossRef]
34. Alimunnisa, J.; Ravichandran, K.; Meena, K. Synthesis and characterization of $Ag@SiO_2$ core-shell nanoparticles for antibacterial and environmental applications. *J. Mol. Liquids* **2017**, *231*, 281–287. [CrossRef]
35. Wei, S.Q.; Yin, X.H.; Li, H.Y.; Du, X.Y.; Zhang, L.M. Multi-Color Fluorescent Carbon Dots: Graphitized Sp^2 Conjugated Domains and Surface State Energy Level Co-Modulate Band Gap Rather Than Size Effects. *Chem. Eur. J.* **2020**, *26*, 8129–8136. [CrossRef]

Disclaimer/Publisher's Note: The statements, opinions and data contained in all publications are solely those of the individual author(s) and contributor(s) and not of MDPI and/or the editor(s). MDPI and/or the editor(s) disclaim responsibility for any injury to people or property resulting from any ideas, methods, instructions or products referred to in the content.

Article

Quantum Yield Enhancement of Carbon Quantum Dots Using Chemical-Free Precursors for Sensing Cr (VI) Ions

Karthiga Anpalagan [1], Hong Yin [2,*], Ivan Cole [2], Tian Zhang [3] and Daniel T. H. Lai [1]

1. Institute of Sustainable Industries and Livable Cities (ISILC), Victoria University, Melbourne, VIC 3011, Australia; karthiga.anpalagan@vu.edu.au (K.A.); daniel.lai@vu.edu.au (D.T.H.L.)
2. School of Engineering, RMIT University, Melbourne, VIC 3000, Australia; ivan.cole@rmit.edu.au
3. Department of Chemical and Biological Engineering, Monash University, Clayton, VIC 3800, Australia; tian.zhang1@monash.edu
* Correspondence: hong.yin@rmit.edu.au; Tel.: +61-3-9925-8259

Abstract: Quantum yield illustrates the efficiency that a fluorophore converts the excitation light into fluorescence emission. The quantum yield of carbon quantum dots (CQDs) can be altered via precursors, fabrication conditions, chemical doping, and surface modifications. In this study, CQDs were first fabricated from whole-meal bread using a chemical-free hydrothermal route, and a low quantum yield (0.81%) was obtained. The combination of whole-meal bread, soybean flour, and lemon juice generated CQDs with almost four folds of enhancement in quantum yield. Detailed characterization suggested that these CQDs were subjected to more complete hydrothermal reactions and had zwitterionic surfaces. The CQDs could selectively detect Cr (VI) ions with a limit of detection (LOD) of 8 ppm. This study shows that the enhancement of the quantum yield of CQDs does not need chemicals, and it is achievable with food precursors.

Keywords: carbon quantum dots; quantum yield; green routes; natural additives; sensing

Citation: Anpalagan, K.; Yin, H.; Cole, I.; Zhang, T.; Lai, D.T.H. Quantum Yield Enhancement of Carbon Quantum Dots Using Chemical-Free Precursors for Sensing Cr (VI) Ions. Inorganics 2024, 12, 96. https://doi.org/10.3390/inorganics12040096

Academic Editor: Ana De Bettencourt-Dias

Received: 9 February 2024
Revised: 14 March 2024
Accepted: 25 March 2024
Published: 28 March 2024

Copyright: © 2024 by the authors. Licensee MDPI, Basel, Switzerland. This article is an open access article distributed under the terms and conditions of the Creative Commons Attribution (CC BY) license (https://creativecommons.org/licenses/by/4.0/).

1. Introduction

Clean water is crucial for human health and well-being, and water pollution leads to illness and even deaths. Any contamination of water with chemicals or other hazardous substances is called water pollution. Nowadays, urbanization is one of the main causes of water pollution. Industries involving dyes, wood preservation, leather tanning, and chrome plating release heavy metal ions into the environment [1,2]. Chromium (VI) is the most common pollutant among many possibly toxic trace elements in industrial wastewater [3–5]. Chromium pollution can cause liver and kidney damage; however, the risk depends on the dose and duration of exposure [6]. Continuous exposure to chromium, even at low concentrations, can damage the skin, eyes, blood, and respiratory and immune systems [7,8]. Moreover, on a cellular level, the genotoxic effect of chromium can lead to oxidative stress, DNA damage, and tumor development [9,10]. Identifying polluted water is an essential measure to prevent Cr exposure and its consequences.

Several analytical techniques, such as atomic absorption spectrometry (AAS), inductively coupled plasma optical mass spectrometry (ICP-MS), and X-ray fluorescence spectroscopy (XRF) are used for the determination of Cr (VI). However, these techniques require bulky, non-portable, and expensive instruments, highly skilled operators, and long analysis times. In contrast, the use of fluorescent sensing materials can achieve fast analysis, low cost, portability, and high sensitivity. Carbon quantum dots (CQDs) are promising candidates for sensing Cr (VI) [11–14]. These fluorescent carbon nanoparticles can be synthesized via various simple methods using sustainable precursors. They possess remarkable optical, physical, biological, and catalytic properties, making them valuable for applications in environmental treatment, analytical techniques, sensing, biomedical devices, and energy regenerations. Quantum confinement effect, surface state and molecule state are three prevailing mechanisms to explain the photoluminescence of CQDs. Quantum confinement splits the energy levels and creates size-dependent bandgaps. When this nanomaterial is illuminated by a photon with sufficient energy, an electron will be excited and move from the valence band to the conduction band, creating an electron-hole pair termed an exciton [15]. When this exciton recombines, the energy will be released as fluorescence. For the surface-state mechanism, functional groups on CQD surfaces have various energy levels and result in a series of emissive traps. A high level of surface oxidation or other effective modifications can increase surface defects, altering emission features of the CQDs [16]. For the molecular state mechanism, the photoluminescence center is an organic fluorophore connected on the surface or interior of the carbon backbone, which exhibits emissions directly [17]. This tunable photoluminescence, along with alterable quantum yield and versatile surface functionalization routes, are some of the major appealing characteristics of CQDs. These features allow a diverse range of sensing applications for CQDs. The quantum yield (QY) of CQD is a substantial feature for fluorometric sensing, and it is defined as the ratio of the number of photons emitted to the number of photons absorbed [18] by the nanomaterial. The variation in precursors, fabrication routes, additives, and doping techniques could change the QY of CQDs.

Most CQDs derived from chemical precursors show comparatively high QYs. For example, anthracite and dimethylformamide produced CQDs with a QY of 47% [19]. In another instance, 80% of QY was obtained for the CQDs derived from citric acid and ethylenediamine [20]. Chang et al. produced CQDs from o-phenylenediamine that exhibited 68% of QY [21]. Generally, CQDs obtained from natural resources have very low QYs because of their complex composition and the difficulty of eliminating impurities. Many efforts have been made to improve the QY, and heteroatom doping and surface modification are the two most effective methods [22,23]. Liu et al. reported that the QY of CQDs from fresh tomatoes was 1.77%, and when they were modified with EDA and urea, the QY increased to 7.9% and 8.5%, respectively [24]. A few chemicals are commonly used to enhance the QY of CQDs. The chemicals with amine groups, such as ethylenediamine, are used as precursors for nitrogen dopants [25] to obtain the expected QY. Likewise, the usage of acids also produced CQDs with higher yields. Phosphoric acid, nitric acid, and citric acid are generally included during the synthesis or after the synthesis as surface passivation agents [26] to enhance the QY. However, when the CQDs derived from natural precursors are treated with chemicals to improve the QY, their biocompatibility is curtailed. These processes contested their safety in many applications.

In our previous research [27], CQDs were fabricated from three types of breads (white bread, whole-meal bread, and multigrain bread) using hydrothermal routes for bioimaging. All obtained CQDs were fluorescent in the absence of any additional surface modification techniques or doping materials. The CQDs prepared from whole-meal bread possessed higher QY (0.81%) than those derived from the other two types of precursors (0.33% and 0.63%, respectively). To further enhance the QY of the bread-derived CQDs without introducing any chemicals, we employed amine-rich soybean flour as a nitrogen source and concentrated lemon juice as an acidic medium for surface modification. Both soybean flour and concentrated lemon juice are easily accessible and cheap, showing great potential to replace harsh chemicals for producing biocompatible CQDs. The obtained chemical-free CQDs were extensively characterized to study the QY enhancement mechanism. The correlations between the physicochemical properties of the CQDs and their QYs were investigated. This study aimed to derive CQDs from edible precursors with enhanced QY and use them to selectively sense chromium (VI) ions.

2. Results and Discussion

2.1. Fabrication of CQD Samples

This study aims to produce CQDs with an enhanced QY using natural resources. Soybean flour and lemon juice are relevant for this purpose. Soybean flour contains more than 50% protein (56% on a moisture-free basis) [28]. Lemon juice is a natural acid source with 1.44 g of citric acid per ounce [29]. The CQDs were prepared by hydrothermal methods, which were very convenient for adding two or more precursors, and the process can be repeated as required. The reproducibility of the synthesis was good, and consistent characterization and sensing results were obtained from CQDs from different batches.

2.2. Characterization Results of the CQDs

Fluorescence QY is a characteristic property that is used to determine the brightness of the nanomaterial. This QY can be measured using two methods: the relative method and the absolute method. In a relative method, the QY is calculated by comparing its fluorescence intensity with another sample with a known QY. For this process, a conventional fluorescence spectrometer with a standard cell holder is sufficient [30]. However, a fluorescence spectrophotometer with an integrated sphere is essential to obtain the absolute QY. This integrating sphere allows the instrument to calculate the number of photons emitted by the sample. Then, the QY is calculated using a comparison between the total number of emitted photons and the total number of absorbed photons. In this method, the QY can be calculated using a single measurement without the need of a reference or absorbance data [31,32]. This absolute method is used in this study to gather QYs for the three samples.

As shown in Table 1, the absolute QY of CQD W derived from whole-meal bread was low (0.81%). Mixing soybean flour with whole-meal bread generated a higher QY (1.42%) in CQD WS. Then, the yield was further enhanced when concentrated lemon juice was used instead of Milli-Q water during the fabrication. The sample derived from whole-meal bread, soybean flour, and lemon juice (CQD WSL) exhibited an improved QY of 2.31%. To understand the reason for the enhanced QY, CQD W, CQD WS, and CQD WSL were extensively characterized.

Table 1. QY measurement of CQD W, CQD WS, and CQD WSL.

Sample Name	Precursors	Absolute QY (%)
CQD W	Whole-meal bread (5 g)	0.81
CQD WS	Whole-meal bread (1 g) + soybean flour (4 g)	1.42
CQD WSL	Whole-meal bread (1 g) + soybean flour (4 g) + lemon juice (30 mL)	2.31

The morphology and size of the CQDs were observed using TEM and the images are shown in Figure 1. The nanoparticles of all three samples look nearly identical. Despite the different precursors, all three CQDs had the same morphology, as they were produced under the same physical conditions. However, the CQD WSL tended to agglomerate more significantly than the other two samples. Tang et al. reported higher photoluminescence efficiency of nanoparticles in the aggregated state than in solution [33]. This enhancement occurred where aggregation limited intramolecular rotation and intramolecular vibration, or highly twisted aggregated structures weakened intermolecular π–π stacking [34]. The aggregation-induced emission (AIE) was also reported for various carbon dots [35], and this could be related to the high QY of CQD WSL.

The XRD patterns of the CQD W, CQD WS, and CQD WSL are shown in Figure 2. All CQDs had a broad peak, suggesting their amorphous structures. Unlike CQD WSL, CQD W and CQD WS also displayed diffraction peaks at 2θ = 44, 64, and 81 degrees. The crystalline structure might be related to the small components from whole meal bread, such as dietary fiber [36], salt or nutrients containing Ca. With the addition of soybean flour, the intensity of these sharp peaks decreased; lemon juice could completely remove the crystalline impurities [37]. In addition, the peak of the amorphous carbon was further broadened by introducing soybean flour and lemon juice, suggesting that these additional precursors promoted the hydrothermal reaction.

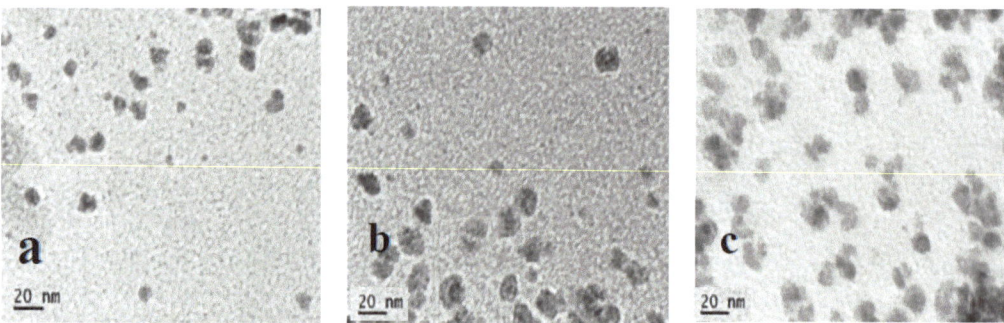

Figure 1. TEM images of (**a**) CQD W, (**b**) CQD WS, and (**c**) CQD WSL.

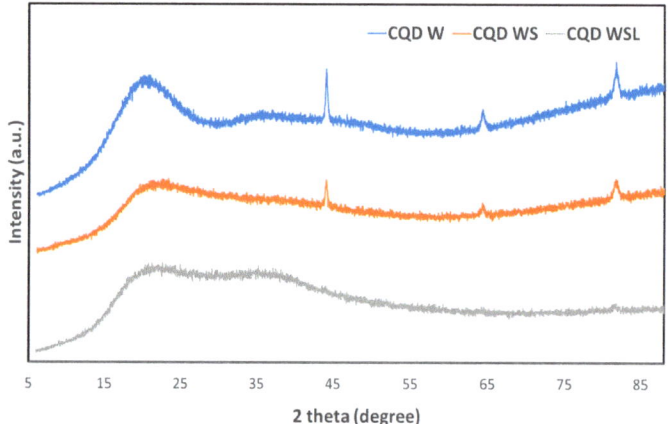

Figure 2. X-ray diffraction patterns of CQD W, CQD WS, and CQD WSL.

FTIR spectroscopy was used to determine the functional groups attached to the CQDs derived from different precursors. The FTIR spectra of CQD W, CQD WS, and CQD WSL are shown in Figure 3. In the fingerprint region, the strong peaks around 850–1150 cm^{-1} corresponded to the C-O groups in the bread. The intensity of these peaks decreased in the order of CQD W > CQD WS > CQD WSL, suggesting that the carbohydrates were decomposed more significantly in the presence of soybean flour and lemon juice. The broad band at 3265 cm^{-1} was related to the stretching vibrations of the hydroxyl group. The shape and position of this peak changed for CQD WS and CQD WSL, implying the presence of NH groups in addition to the -OH groups. The stretching vibration at 1650 cm^{-1} was related to the amide group, indicating that proteins were involved in the hydrothermal process. The peak at 1150 cm^{-1} attributed to the C-N bond was found in CQD W and decreased in CQD WS, while this peak disappeared in CQD WSL. Instead, a new peak at 1190 emerged, which might be the C-N bond generated under the action of lemon juice. The C=O stretching at 1701 cm^{-1} was only identified in the CQD WSL, related to the acid added as a precursor. The co-presence of amide and carboxylic groups in CQD WLS might lead to a zwitterionic surface, which will be further discussed in the zeta potential results.

XPS was used to analyze the surface chemistry of the CQDs. The XPS results indicated that all three CQDs mainly consisted of C, O, and N. The atomic compositions of the CQD surface are shown in Table 2. N was successfully doped in the CQDs. The energy states of CQDs changed with N-doping, and the new surface-emissive trap states created by N could promote the radiative recombination of excitons [38–40]. The photoluminescence of N-doped CQDs emerged from the radiative recombination of photo-induced electrons in the surface trap states, and holes at the highest occupied molecular orbitals (HOMO) [41,42]. Due to this, CQDs with the least N doping (CQD W) had the lowest QY. However, CQD WS with the highest N content did not show the highest QY, suggesting that the QY was not solely dependent on N-doping.

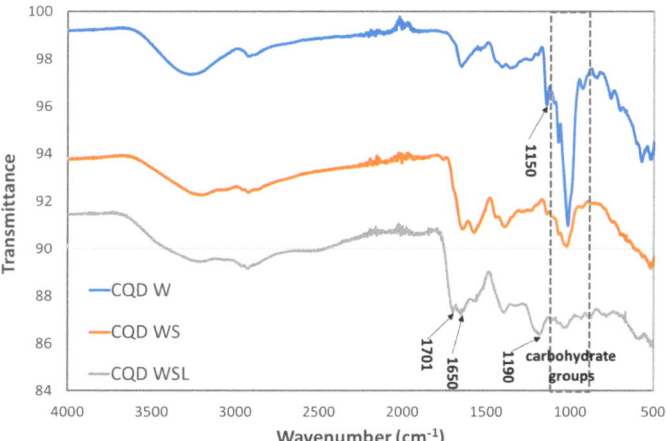

Figure 3. FTIR spectra of CQD W, CQD WS, and CQD WSL.

Table 2. Atomic composition of CQD W, CQD WS, and CQD WSL.

CQD	O (%)	C (%)	N (%)
CQD W	56.76	37.91	5.32
CQD WS	49.76	39.35	10.88
CQD WSL	52.91	32.99	7.87

The high-resolution C 1s spectra are shown in Figure 4. The C peak was deconvoluted into three peaks at 288.32 eV, 283.9 eV, and 286 eV, attributed to C=O, C-O/C-N, and C-C bonds, respectively [43]. The peak related to C=O was more distinct in the CQD WSL, and the ratio between C=O and C-O/C-N increased from CQD W to CQD WS to CQD WSL, suggesting that C-O was gradually oxidized to form ester bonds or carboxylic bonds. It was reported that strong oxidizing acids carbonized small organic molecules to carbonaceous materials, which could be further cut into small sheets by controlled oxidation [44]. As shown in Figure 5, O 1s spectra exhibited two peaks at around 531 eV and 533 eV, related to the C=O and C-O bonds, respectively. The C=O peak gradually grew, and the ratio between C=O/C-O increased significantly for CQD WSL. It was consistent with the C 1s spectra and confirmed that lemon juice promoted the hydrothermal reaction of carbohydrates and achieved a more complete reaction towards small active carbon materials with a high QY.

Figure 6 illustrates the pH dependence of zeta potential. CQD WSL had the smallest absolute values of zeta potential with the increasing pH. The small surface charge might induce more aggregation because of the weaker repulsive force between individual particles to separate them. This observation is consistent with the TEM image of CQD WSL. The pH-independent low (absolute) zeta potential also suggested that the CQD WSL had a zwitterionic surface, which is characterized by the presence of equal amounts of cationic and anionic groups and a net surface charge of zero. Radchanka et al. reported that the net charge density within the slipping plane around the quantum dots affected the nonradiative recombination processes. As a general trend, QDs with zwitterionic surface groups and low surface charges demonstrated a high QY due to effective passivation from surface-related nonradiative processes [45].

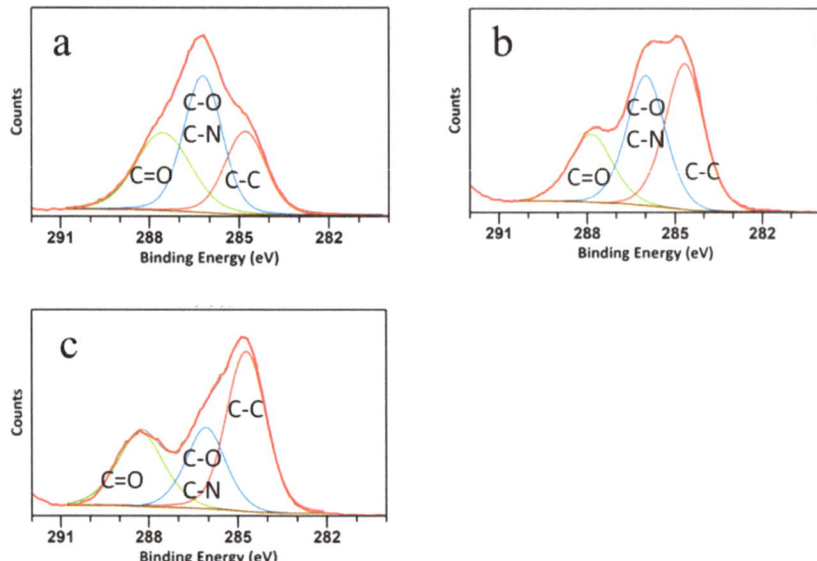

Figure 4. C 1s spectra of (**a**) CQD W, (**b**) CQD WS, and (**c**) CQD WSL. Lines in different colors represent the fitting results of C=O (green), C-O and C-N (blue), and C-C (maroon).

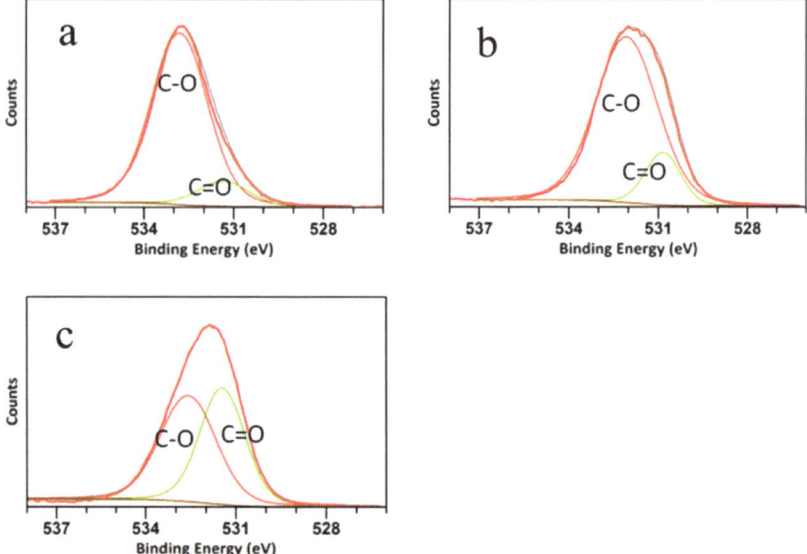

Figure 5. O 1s spectra of (**a**) CQD W, (**b**) CQD WS, and (**c**) CQD WSL.

According to the characterization results, nitrogen was doped in the CQDs by soybean flour. The acid in lemon juice further promoted the hydrothermal reaction. The CQD WSL derived from whole-meal bread + soybean flour + lemon juice had the highest QY (2.31%). This high QY might be due to its complete reaction promoted by the lemon juice, and its zwitterionic surface containing carboxylic acid group (from lemon) and amide groups (from soy flour).

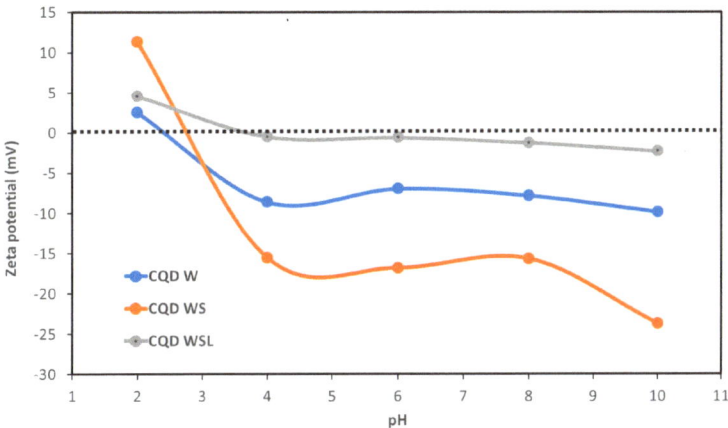

Figure 6. pH-dependent zeta potentials of CQD W, CQD WS, and CQD WSL.

Figure 7a shows the fluorescence spectra of CQD W, CQD WS, and CQD WSL. All spectra exhibited a broad peak from 400 to 500 nm. The peak intensity followed the order of CQD W < CQD WS < CQD WSL, consistent with the trend of QY. The CQDs were used to sense various heavy metal ions in aqueous solutions. As shown in Figure 7b–d, the fluorescence intensity changed insignificantly in the presence of nine potentially toxic ions, except for Cr (VI) ions. Adding 100 ppm of Cr (VI) could quench the fluorescence of CQD W, CQD WS, and CQD WSL, showing their unique responses to Cr ions. The CQD WSL sample showed the most significant fluorescence quenching in terms of intensity loss, i.e., $(F_0 - F)/F_0$, where F_0 and F are the fluorescence of CQD WSL in the absence and presence of 100 ppm Cr (VI) ions. As shown in Figure 7e, the fluorescence of CQD WSL decreased with the increasing concentration of Cr (VI) ions. The fluorescence quenching followed a linear relationship from 2.5 to 50 ppm (Figure 7f). The limit of detection (LOD) was estimated as 8 ppm based on 3σ/slope. The LOD of the CQD WSL derived from 100% edible precursors was significantly lower than the LOD of gold nano-double cone @ silver nanorods (1.69 μM, i.e., 80 ppm). Table 3 compares the Cr (VI) sensing performance (LOD and linear range) of reported CQDs and the CQDs synthesized in this study. Considering that the lowest permissible level in dischargeable water set by the World Health Organization (WHO) is 1 ppm, future work will need to focus on reducing the LOD by enhancing the fluorescence QY, potentially through introducing more N-containing precursors. In addition, the sensing selectivity of CQD WSL was assessed by measuring the fluorescence quenching of CQD WSL in the presence of 100 ppm of various heavy metal ions (Figure 7g) and in the co-presence of 100 ppm of Cr (VI) and 100 ppm of other heavy metal ions. (Figure 7h). The quenching remained high when Cr (VI) ions co-existed with various interfering heavy metal ions, suggesting the high selectivity of the CQD WSL sensor.

Table 3. Comparison of CQDs used for Cr (VI) sensing.

Precursors	LOD	Linearity Range	Reference
Citric acid and glycine	4.16 μmol L^{-1}	5 to 200 μmol L^{-1}	[14]
Fullerene, H_2O_2, and NH_4OH	300 nM	1–100 μM	[13]
Poria cocos polysaccharide	0.25 μM	1–100 μM	[12]
Ammonium citrate and bis(pinacolato) diboron	0.24 μM	0.3–500 μM	[46]
Diacetone acrylamide and 3-aminopropyltriethoxysilane	0.995 μM	0–200 μM	[47]
Whole-meal bread, soybean flour, and lemon juice	8 ppm	2.5–50 ppm	This article

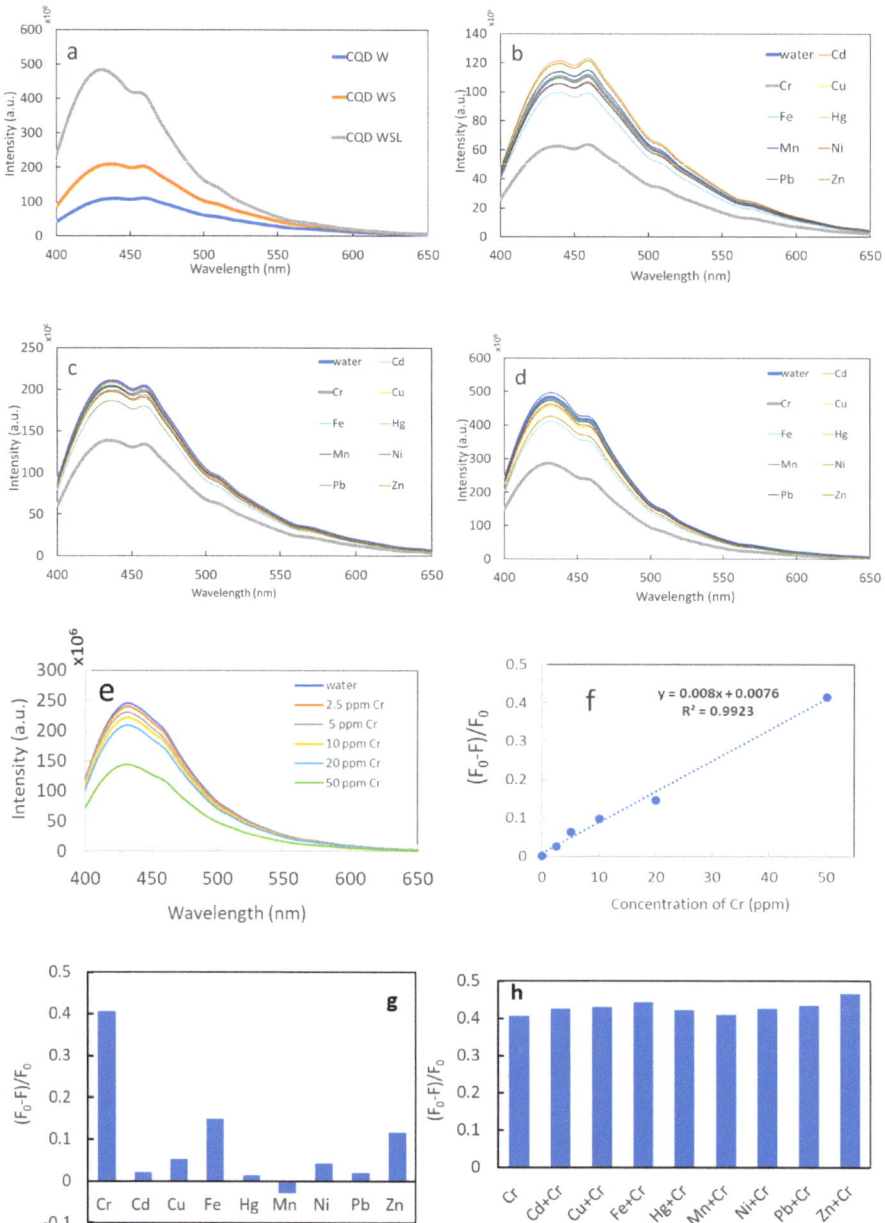

Figure 7. Fluorescence spectra of (**a**) CQD W, CQD WS, and CQD WSL; (**b**) CQD W after adding various heavy metal ion solutions (100 ppm), (**c**) CQD WS after adding various heavy metal ion solutions (100 ppm), and (**d**) CQD WSL after adding various heavy metal ion solutions (100 ppm). (**e**) Fluorescence spectra of CQD WSL in presence of Cr solutions of various concentrations, (**f**) fluorescence quenching of CQD WSL vs. Cr concentration, (**g**) fluorescence quenching of CQD WSL in the presence of 100 ppm of various heavy metal ions, and (**h**) fluorescence quenching of CQD WSL in the presence of 100 ppm of Cr (VI) and 100 ppm of other heavy metal ions.

3. Materials and Methods

3.1. Fabrication of CQDs

Firstly, the CQDs were fabricated from whole-meal bread. Every serving of the whole-meal bread (approximately 35 g) contained 3.86 g of protein, 2.48 g of fat, 23.64 g of carbohydrates, 2.8 g of fiber, and 0.3 g of sodium and potassium. After analyzing its QY, soybean flour was mixed with whole-meal bread in the second attempt. Lemon juice was also included in the next attempt to further enhance the QY. The samples were labeled in relation to the type of precursors (Table 4).

Table 4. Details of the CQDs prepared from different precursors.

Sample Name	Precursors
CQD W	Whole-meal bread (5 g)
CQD WS	Whole-meal bread (1 g) + soybean flour (4 g)
CQD WSL	Whole-meal bread (1 g) + soybean flour (4 g) + lemon juice (30 mL)

To prepare CQD W, 5 g (±0.1 mg) of whole-meal bread (purchased from the local supermarket) was dispersed in 30 mL of Milli-Q water by sonicating (Soniclean, LABOUIP Technologies, Bayswater, Australia) for 5 min and then the solution was transferred into a 50 mL autoclave chamber (Robotdigg equip makers, HK, China). This chamber was heated at 180 °C for 4 h in an oven (RHTOV2HP, Russell Hobbs, Spectrum brands, Victoria, Australia). After heating, the chamber was allowed to cool to room temperature overnight. The resulting sample was transferred into a centrifuge tube and centrifuged (MSE centrifuge, Thomas Scientific, Swedesboro, NJ, USA) for 5 min at 3000 rpm. The collected supernatant was filtered using a number 1 (90 mm) filter paper (Whatman, GE Healthcare UK Limited, Amersham, UK) and a 0.22 µm syringe filter membrane (Millex®, Merck Millipore, Merck KGaA, Darmstadt, Germany), sequentially. The sample was labelled CQD W (whole-meal bread) and stored in a refrigerator at 4 °C.

CQD WS was prepared by dispersing 1 g (±0.1 mg) of whole-meal bread and 4 g (±0.1 mg) of soybean flour (purchased from the local supermarket) in 30 mL of Milli-Q water using sonication for 5 min. The solution was heated at 180 °C for 4 h followed by centrifuge and filtration. The procedure was the same as the preparation of CQD W. For the preparation of CQD WSL, 1 g (±0.1 mg) of whole-meal bread and 4 g (±0.1 mg) of soybean flour were dispersed in 30 mL of freshly squeezed pulp-free lemon juice instead of in Milli-Q water. The precursors were dispersed using sonication for 5 min. The rest of the procedures, including heating, centrifugation, and filtration, were the same as the preparation of CQD W.

3.2. Characterization of CQDs

3.2.1. QY Measurement

Fluorescence QY was obtained on a photoluminescence spectrometer FT300 (PicoQuant GmbH, Berlin, Germany) using an integrated sphere. The quartz cuvette (1 cm × 1 cm × 5 cm) containing the sample solution was positioned in 'IN mode' with 20-degree tilting, where the 423 nm excitation laser directly transmitted through the cuvette. For the measurement of low QY values, an emission attenuator (attenuation level at 100) was applied for less than 445 nm, and the correction factors were included in the QY calculation.

3.2.2. Transmission Electron Microscopy (TEM)

TEM images were captured using a JEOL 1010 TEM 868 (JEOL, Sydney, Australia) operated at an accelerating voltage of 100 kV. The CQDs were sonicated for 20 min, and the resulting supernatant was drop-casted on a carbon grid and dried overnight under ambient conditions for analysis.

3.2.3. X-ray Diffraction (XRD)

XRD patterns were collected on a Bruker AXS D8 Discover diffractometer (Bruker, Melbourne, Australia) equipped with a Cu Kα radiation source (λ = 1.5418 Å) operating at 40 kV and 35 mA.

3.2.4. Fourier-Transform Infrared Spectroscopy (FTIR)

FTIR spectra were recorded using an FTIR spectrometer (PerkinElmer, Waltham, MA, USA) with an average of 16 scans per sample and a resolution of 4 cm^{-1} in the range of 4000–500 cm^{-1}.

3.2.5. X-ray Photoelectron Spectroscopy (XPS)

XPS was performed using a Thermo Scientific K-Alpha XPS system with a monochromate Al 1487 eV Kα source. CasaXPS 2.3.23 software was used for peak fitting and background signal subtraction.

3.2.6. Zeta Potential Measurement

A Malvern Zetasizer Nano Z system was used for zeta-potential measurements. Briefly, 10 mg of each CQD sample was placed in a plastic cuvette containing 3 mL of dispersion medium. The dispersion medium was prepared in advance based on deionized water with the pH (pH = 2, 4, 6, 8, and 10) adjusted by adding 0.1 M HCl or 0.1 M NaOH. The cuvette was placed in an ultrasonic bath for 10 s and then shaken manually to ensure good dispersion. The temperature of all measurements was maintained at 25 °C. The cuvette was thoroughly washed with deionized water before and after each measurement.

3.3. Heavy Metal Ion Sensing

CQD solution (1 mg/mL, 100 µL) was transferred into a 96-well plate with a clear glass bottom. Then, 100 µL of various metal ion solutions with a known concentration was added into each well and mixed thoroughly. The fluorescence spectra before and after adding metal ions were recorded using a SpectraMax Paradigm Plate Reader (Molecular Devices, San Jose, CA, USA) with an excitation wavelength of 360 nm. For measuring the dependence of fluorescence of CQD WSL on the Cr (VI) concentration, CQD WSL solution (100 µL) was transferred into a 96-well plate with a clear glass bottom. Then, 100 µL Cr (VI) solution with a known concentration was added into each well and mixed thoroughly. The fluorescence spectra before and after adding Cr (VI) solution were recorded.

4. Conclusions

This study fabricated CQDs from whole-meal bread and successfully improved their QY using natural and edible additives, such as soy flour and lemon juice. Despite the different precursors, all CQDs had the same morphology as they were produced under the same physical conditions. CQD WSL tended to agglomerate more significantly than the other two samples, possibly because of its neutral surface charge. According to the characterization results, nitrogen was doped in the CQDs. Compared to the QY of CQD W (0.81%), the CQDs derived from whole-meal bread + soybean flour + lemon juice (CQD WSL) had the highest QY (2.31%). FTIR, XRD, and XPS results confirmed that the bread precursors were subjected to more complete hydrothermal reactions in the presence of soy flour and lemon juice, which enhanced the QYs. Evidenced by zeta potential measurements, CQD WSL had a zwitterionic surface, containing a carboxylic acid group (from lemon) and amide groups (from soy flour), which was also responsible for its high QY. The fluorescent CQDs were tested for metal ion sensing, and the emission intensities followed the order of CQD W < CQD WS < CQD WSL, consistent with the trend of QY. CQD WSL was able to sense Cr (VI) selectively with an LOD of 8 ppm. As these CQDs were derived entirely from edible resources, they are extremely safe for sensing applications. Also, the fabrication of CQD is quick and efficient, providing a sustainable method to identify Cr (VI)-polluted water and support health and well-being.

Author Contributions: Methodology, K.A., H.Y. and D.T.H.L.; Resources, H.Y. and T.Z.; Supervision, I.C. and D.T.H.L.; Writing—original draft, K.A. and H.Y.; Writing—review and editing, K.A., H.Y., I.C. and D.T.H.L. All authors have read and agreed to the published version of the manuscript.

Funding: This work was possible through the funding support of a Defense Science Institute (DSI) Collaborative Research Grant (#CR-0030) and RMIT Innovation Proof of Concept Fund 2022 (TIS00066).

Data Availability Statement: The raw data supporting the conclusions of this article will be made available by the authors on request.

Acknowledgments: The authors acknowledge the facilities and the scientific and technical assistance of the RMIT Microscopy & Microanalysis Facility (RMMF), a linked laboratory of Microscopy Australia and the RMIT Micro Nano Research Facility (MNRF).

Conflicts of Interest: The authors declare that the research was conducted in the absence of any commercial or financial relationships that could be construed as potential conflicts of interest. The funders had no role in the design of the study; in the collection, analyses, or interpretation of data; in the writing of the manuscript; or in the decision to publish the results.

References

1. International Agency for Research on Cancer. *IARC Monographs on the Evaluation of Carcinogenic Risks to Humans: Occupational Exposures in Insecticide Application, and Some Pesticides*; IARC: Lyon, France, 1991; Volume 53.
2. Wilbur, S.B. *Toxicological Profile for Chromium*; Agency for Toxic Substances and Disease Registry: Atlanta, GA, USA, 2000.
3. Shanker, A.; Venkateswarlu, B.; Nriagu, J. *Encyclopedia of Environmental Health*; Elsevier: Amsterdam, The Netherlands, 2011.
4. Yan, B.-Z.; Chen, Z.-F. Influence of pH on Cr(VI) reduction by organic reducing substances from sugarcane molasses. *Appl. Water Sci.* **2019**, *9*, 61. [CrossRef]
5. Wu, M.; Li, G.; Jiang, X.; Xiao, Q.; Niu, M.; Wang, Z.; Wang, Y. Non-biological reduction of Cr(vi) by reacting with humic acids composted from cattle manure. *RSC Adv.* **2017**, *7*, 26903–26911. [CrossRef]
6. Pellerin, C.; Booker, S.M. Reflections on hexavalent chromium: Health hazards of an industrial heavyweight. *Environ. Health Perspect.* **2000**, *108*, A402–A407. [CrossRef] [PubMed]
7. Zhang, R.; Xiang, Y.; Ran, Q.; Deng, X.; Xiao, Y.; Xiang, L.; Li, Z. Involvement of Calcium, Reactive Oxygen Species, and ATP in Hexavalent Chromium-Induced Damage in Red Blood Cells. *Cell. Physiol. Biochem.* **2014**, *34*, 1780–1791. [CrossRef]
8. Wilbur, S.; Abadin, H.; Fay, M.; Yu, D.; Tencza, B.; Ingerman, L.; Klotzbach, J.; James, S. Health effects. In *Toxicological Profile for Chromium*; Agency for Toxic Substances and Disease Registry (US): Atlanta, GA, USA, 2012.
9. Wise, J.T.; Wang, L.; Xu, J.; Zhang, Z.; Shi, X. Oxidative stress of Cr (III) and carcinogenesis. In *The Nutritional Biochemistry of Chromium (III)*; Elsevier: Amsterdam, The Netherlands, 2019; pp. 323–340.
10. Mishra, S.; Bharagava, R.N. Toxic and genotoxic effects of hexavalent chromium in environment and its bioremediation strategies. *J. Environ. Sci. Health Part C* **2015**, *34*, 1–32. [CrossRef] [PubMed]
11. Nghia, N.N.; Huy, B.T.; Lee, Y.-I. Colorimetric detection of chromium(VI) using graphene oxide nanoparticles acting as a peroxidase mimetic catalyst and 8-hydroxyquinoline as an inhibitor. *Microchim. Acta* **2018**, *186*, 36. [CrossRef]
12. Huang, Q.; Bao, Q.; Wu, C.; Hu, M.; Chen, Y.; Wang, L.; Chen, W. Carbon dots derived from Poria cocos polysaccharide as an effective "on-off" fluorescence sensor for chromium (VI) detection. *J. Pharm. Anal.* **2021**, *12*, 104–112. [CrossRef]
13. Babazadeh, S.; Bisauriya, R.; Carbone, M.; Roselli, L.; Cecchetti, D.; Bauer, E.M.; Sennato, S.; Prosposito, P.; Pizzoferrato, R. Colorimetric Detection of Chromium(VI) Ions in Water Using Unfolded-Fullerene Carbon Nanoparticles. *Sensors* **2021**, *21*, 6353. [CrossRef] [PubMed]
14. Wang, H.; Liu, S.; Xie, Y.; Bi, J.; Li, Y.; Song, Y.; Cheng, S.; Li, D.; Tan, M. Facile one-step synthesis of highly luminescent N-doped carbon dots as an efficient fluorescent probe for chromium(vi) detection based on the inner filter effect. *New J. Chem.* **2018**, *42*, 3729–3735. [CrossRef]
15. Zhang, Z.; Sung, J.; Toolan, D.T.W.; Han, S.; Pandya, R.; Weir, M.P.; Xiao, J.; Dowland, S.; Liu, M.; Ryan, A.J.; et al. Ultrafast exciton transport at early times in quantum dot solids. *Nat. Mater.* **2022**, *21*, 533–539. [CrossRef]
16. Ding, H.; Yu, S.-B.; Wei, J.-S.; Xiong, H.-M. Full-Color Light-Emitting Carbon Dots with a Surface-State-Controlled Luminescence Mechanism. *ACS Nano* **2015**, *10*, 484–491. [CrossRef] [PubMed]
17. Krysmann, M.J.; Kelarakis, A.; Dallas, P.; Giannelis, E.P. Formation Mechanism of Carbogenic Nanoparticles with Dual Photoluminescence Emission. *J. Am. Chem. Soc.* **2011**, *134*, 747–750. [CrossRef] [PubMed]
18. Sommer, M.E.; Elgeti, M.; Hildebrand, P.W.; Szczepek, M.; Hofmann, K.P.; Scheerer, P. Structure-based biophysical analysis of the interaction of rhodopsin with G protein and arrestin. In *Methods in Enzymology*; Elsevier: Amsterdam, The Netherlands, 2015; pp. 563–608.
19. Li, M.; Yu, C.; Hu, C.; Yang, W.; Zhao, C.; Wang, S.; Zhang, M.; Zhao, J.; Wang, X.; Qiu, J. Solvothermal conversion of coal into nitrogen-doped carbon dots with singlet oxygen generation and high quantum yield. *Chem. Eng. J.* **2017**, *320*, 570–575. [CrossRef]

20. Zhu, S.; Meng, Q.; Wang, L.; Zhang, J.; Song, Y.; Jin, H.; Zhang, K.; Sun, H.; Wang, H.; Yang, B. Highly Photoluminescent Carbon Dots for Multicolor Patterning, Sensors, and Bioimaging. *Angew. Chem. Int. Ed.* **2013**, *52*, 4045–4049. [CrossRef]
21. Chang, C.-Y.; Venkatesan, S.; Herman, A.; Wang, C.-L.; Teng, H.; Lee, Y.-L. Carbon quantum dots with high quantum yield prepared by heterogeneous nucleation processes. *J. Alloys Compd.* **2023**, *938*, 168654. [CrossRef]
22. Lou, Y.; Hao, X.; Liao, L.; Zhang, K.; Chen, S.; Li, Z.; Ou, J.; Qin, A.; Li, Z. Recent advances of biomass carbon dots on syntheses, characterization, luminescence mechanism, and sensing applications. *Nano Sel.* **2021**, *2*, 1117–1145. [CrossRef]
23. de Oliveira, B.P.; da Silva Abreu, F.O.M. Carbon quantum dots synthesis from waste and by-products: Perspectives and challenges. *Mater. Lett.* **2021**, *282*, 128764. [CrossRef]
24. Liu, W.; Li, C.; Sun, X.; Pan, W.; Yu, G.; Wang, J. Highly crystalline carbon dots from fresh tomato: UV emission and quantum confinement. *Nanotechnology* **2017**, *28*, 485705. [CrossRef] [PubMed]
25. Zheng, J.; Xie, Y.; Wei, Y.; Yang, Y.; Liu, X.; Chen, Y.; Xu, B. An Efficient Synthesis and Photoelectric Properties of Green Carbon Quantum Dots with High Fluorescent Quantum Yield. *Nanomaterials* **2020**, *10*, 82. [CrossRef]
26. Wu, M.; Wang, Y.; Wu, W.; Hu, C.; Wang, X.; Zheng, J.; Li, Z.; Jiang, B.; Qiu, J. Preparation of functionalized water-soluble photoluminescent carbon quantum dots from petroleum coke. *Carbon* **2014**, *78*, 480–489. [CrossRef]
27. Anpalagan, K.; Karakkat, J.V.; Jelinek, R.; Kadamannil, N.N.; Zhang, T.; Cole, I.; Nurgali, K.; Yin, H.; Lai, D.T.H. A Green Synthesis Route to Derive Carbon Quantum Dots for Bioimaging Cancer Cells. *Nanomaterials* **2023**, *13*, 2103. [CrossRef] [PubMed]
28. Porter, M.A.; Jones, A.M. Variability in soy flour composition. *J. Am. Oil Chem. Soc.* **2003**, *80*, 557–562. [CrossRef]
29. Penniston, K.L.; Nakada, S.Y.; Holmes, R.P.; Assimos, D.G. Quantitative Assessment of Citric Acid in Lemon Juice, Lime Juice, and Commercially-Available Fruit Juice Products. *J. Endourol.* **2008**, *22*, 567–570. [CrossRef] [PubMed]
30. Würth, C.; Grabolle, M.; Pauli, J.; Spieles, M.; Resch-Genger, U. Relative and absolute determination of fluorescence quantum yields of transparent samples. *Nat. Protoc.* **2013**, *8*, 1535–1550. [CrossRef] [PubMed]
31. Faulkner, D.O.; McDowell, J.J.; Price, A.J.; Perovic, D.D.; Kherani, N.P.; Ozin, G.A. Measurement of absolute photoluminescence quantum yields using integrating spheres—Which way to go? *Laser Photonics Rev.* **2012**, *6*, 802–806. [CrossRef]
32. Porrès, L.; Holland, A.; Pålsson, L.-O.; Monkman, A.P.; Kemp, C.; Beeby, A. Absolute Measurements of Photoluminescence Quantum Yields of Solutions Using an Integrating Sphere. *J. Fluoresc.* **2006**, *16*, 267–273. [CrossRef]
33. Luo, J.; Xie, Z.; Xie, Z.; Lam, J.W.Y.; Cheng, L.; Chen, H.; Qiu, C.; Kwok, H.S.; Zhan, X.; Liu, Y.; et al. Aggregation-Induced Emission of 1-Methyl-1,2,3,4,5-Pentaphenylsilole. *Chem. Commun.* **2001**, *18*, 1740–1741. [CrossRef] [PubMed]
34. Zhang, H.; Zhao, Z.; Turley, A.T.; Wang, L.; McGonigal, P.R.; Tu, Y.; Li, Y.; Wang, Z.; Kwok, R.T.K.; Lam, J.W.Y.; et al. Aggregate Science: From Structures to Properties. *Adv. Mater.* **2020**, *32*, e2001457. [CrossRef] [PubMed]
35. Gao, M.X.; Liu, C.F.; Wu, Z.; Zeng, Q.L.; Yang, X.X.; Wu, W.B.; Li, Y.F.; Huang, C.Z. A surfactant-assisted redox hydrothermal route to prepare highly photoluminescent carbon quantum dots with aggregation-induced emission enhancement properties. *Chem. Commun.* **2013**, *49*, 8015–8017. [CrossRef]
36. Kaur, B.; Panesar, P.S.; Thakur, A. Extraction and evaluation of structural and physicochemical properties of dietary fiber concentrate from mango peels by using green approach. *Biomass-Convers. Biorefin.* **2021**, 1–10. [CrossRef]
37. Chaudhary, S.; Kumar, S.; Kaur, B.; Mehta, S.K. Potential prospects for carbon dots as a fluorescence sensing probe for metal ions. *RSC Adv.* **2016**, *6*, 90526–90536. [CrossRef]
38. Tang, L.; Ji, R.; Cao, X.; Lin, J.; Jiang, H.; Li, X.; Teng, K.S.; Luk, C.M.; Zeng, S.; Hao, J.; et al. Deep Ultraviolet Photoluminescence of Water-Soluble Self-Passivated Graphene Quantum Dots. *ACS Nano* **2012**, *6*, 5102–5110. [CrossRef]
39. Kwon, W.; Do, S.; Lee, J.; Hwang, S.; Kim, J.K.; Rhee, S.-W. Freestanding Luminescent Films of Nitrogen-Rich Carbon Nanodots toward Large-Scale Phosphor-Based White-Light-Emitting Devices. *Chem. Mater.* **2013**, *25*, 1893–1899. [CrossRef]
40. Liu, Q.; Guo, B.; Rao, Z.; Zhang, B.; Gong, J.R. Strong Two-Photon-Induced Fluorescence from Photostable, Biocompatible Nitrogen-Doped Graphene Quantum Dots for Cellular and Deep-Tissue Imaging. *Nano Lett.* **2013**, *13*, 2436–2441. [CrossRef]
41. Liu, J.; Liu, X.; Luo, H.; Gao, Y. One-step preparation of nitrogen-doped and surface-passivated carbon quantum dots with high quantum yield and excellent optical properties. *RSC Adv.* **2014**, *4*, 7648–7654. [CrossRef]
42. Liao, J.; Cheng, Z.; Zhou, L. Nitrogen-Doping Enhanced Fluorescent Carbon Dots: Green Synthesis and Their Applications for Bioimaging and Label-Free Detection of Au^{3+} Ions. *ACS Sustain. Chem. Eng.* **2016**, *4*, 3053–3061. [CrossRef]
43. Huang, H.; Li, S.; Chen, B.; Wang, Y.; Shen, Z.; Qiu, M.; Pan, H.; Wang, W.; Wang, Y.; Li, X. Endoplasmic reticulum-targeted polymer dots encapsulated with ultrasonic synthesized near-infrared carbon nanodots and their application for in vivo monitoring of Cu^{2+}. *J. Colloid Interface Sci.* **2022**, *627*, 705–715. [CrossRef]
44. Wang, Y.; Hu, A. Carbon quantum dots: Synthesis, properties and applications. *J. Mater. Chem. C* **2014**, *2*, 6921–6939. [CrossRef]
45. Radchanka, A.; Hrybouskaya, V.; Iodchik, A.; Achtstein, A.W.; Artemyev, M. Zeta Potential-Based Control of CdSe/ZnS Quantum Dot Photoluminescence. *J. Phys. Chem. Lett.* **2022**, *13*, 4912–4917. [CrossRef]
46. Wang, Y.; Hu, X.; Li, W.; Huang, X.; Li, Z.; Zhang, W.; Zhang, X.; Zou, X.; Shi, J. Preparation of boron nitrogen co-doped carbon quantum dots for rapid detection of Cr(VI). *Spectrochim. Acta Part A Mol. Biomol. Spectrosc.* **2020**, *243*, 118807. [CrossRef]
47. Zhang, J.; Jing, C.; Wang, B. A label-free fluorescent sensor based on Si, N-codoped carbon quantum dots with enhanced sensitivity for the determination of Cr (VI). *Materials* **2022**, *15*, 1733. [CrossRef] [PubMed]

Disclaimer/Publisher's Note: The statements, opinions and data contained in all publications are solely those of the individual author(s) and contributor(s) and not of MDPI and/or the editor(s). MDPI and/or the editor(s) disclaim responsibility for any injury to people or property resulting from any ideas, methods, instructions or products referred to in the content.

Article

Orange Peel Biochar–CdS Composites for Photocatalytic Hydrogen Production

Xiang Li [1,2,3], Yuxin Zang [2], Jindi Zhang [1], Lili Zhang [1], Jing Zhang [2], Mengyang Huang [1,*] and Jiaqiang Wang [1,2,3,*]

[1] School of Chemistry and Resources Engineering, Honghe University, Mengzi 661100, China; lx1535655417@163.com (X.L.); zhangjindi@uoh.edu.cn (J.Z.); zhanglili@uoh.edu.cn (L.Z.)
[2] School of Materials and Energy, Chemical Sciences & Technology, Institute of International Rivers and Eco-Security, Yunnan University, Kunming 650091, China; 458451215@163.com (Y.Z.); quayjing@163.com (J.Z.)
[3] Institute of Frontier Technologies in Water Treatment Co., Ltd., Kunming 650503, China
* Correspondence: huangmengyang@uoh.edu.cn (M.H.); jqwang@ynu.edu.cn (J.W.)

Abstract: Orange peel biochar (C)-supported cadmium sulfide composites (CdS-C) were prepared by the combination of hydrothermal and calcination methods. The structure and morphology were characterized in detail by X-ray diffraction (XRD) and scanning electron microscopy (SEM), respectively. The CdS-C composite with 60% CdS exhibited the highest photocatalytic hydrogen production rate of 7.8 mmol·g^{-1}·h^{-1}, approximately 3.69 times higher than that of synthesized CdS without biochar. These results indicate that biochar derived from orange peel could be a low-cost, renewable, environmentally friendly, and metal-free co-catalyst for CdS, enhancing its photostability.

Keywords: photocatalytic hydrogen production; biochar; CdS; hydrothermal method

Citation: Li, X.; Zang, Y.; Zhang, J.; Zhang, L.; Zhang, J.; Huang, M.; Wang, J. Orange Peel Biochar–CdS Composites for Photocatalytic Hydrogen Production. *Inorganics* **2024**, *12*, 156. https://doi.org/10.3390/inorganics12060156

Academic Editor: Hicham Idriss

Received: 31 March 2024
Revised: 26 May 2024
Accepted: 28 May 2024
Published: 31 May 2024

Copyright: © 2024 by the authors. Licensee MDPI, Basel, Switzerland. This article is an open access article distributed under the terms and conditions of the Creative Commons Attribution (CC BY) license (https://creativecommons.org/licenses/by/4.0/).

1. Introduction

Among the existing renewable energy options, hydrogen is a clean, efficient, safe, and sustainable secondary energy source. It is currently recognized as the most ideal energy carrier and is considered the most promising new energy source of the 21st century. Solar energy, if effectively collected, converted, and stored, can be the greenest and most abundant energy source on Earth [1]. Among existing technologies, the use of visible light to activate photocatalytic reactions in semiconductor materials is considered one of the most promising and revolutionary approaches for addressing environmental pollution and meeting the challenges posed by global energy shortages [2].

The development of non-precious metal photocatalytic materials has long been an urgent need for fuel production and environmental treatment. Non-metallic carbon materials have proven to be a promising alternative as co-catalysts due to their low cost, ecological compatibility, good electrical conductivity, chemical stability, and photothermal effects [3–5]. Biomass-derived charcoal is an environmentally friendly material that is efficient, cheap, and easy to prepare. It has received widespread attention in the fields of materials science, chemistry, and environmental protection [6–9]. Biochar has a high specific surface area, high porosity, and sufficient surface functional groups. Its excellent ion exchange capacity and high stability make it a suitable material for pollution control [10]. To date, biochar materials from agricultural waste have been introduced for the preparation of composites for the photocatalytic removal of organic pollutants from water as well as for the production of hydrogen through water decomposition, supercapacitors, fuel cells, CO$_2$ adsorption, and energy storage [11–14]. More importantly, the synergistic effect of biochar with other metals or metal oxides can enhance the adsorption capacity of biochar-based catalysts, increase the visible light absorption, improve the separation of photogenerated electrons and holes, and reduce the bandgap, thereby improving the photocatalytic performance of biochar-based catalysts [15–18]. For example, biochar serves as the charge transfer mediator between Ag_3PO_4 and Fe_3O_4 and hinders the recombination of electron–hole pairs (e^--h^+). The efficient separation of e^--h^+ pairs minimizes recombination, thereby boosting charge transfer and separation, which significantly improves the photocatalytic ability of biochar-based catalysts [19]. Modified biochar has been used as a substrate to anchor g–C_3N_4–$FeVO_4$ heterojunctions to extend charge lifetime and enhance the adsorption capacity of the composites. Using a g–C_3N_4–$FeVO_4$-BC ternary system effectively achieved 98.4% removal of methylparaben degradation via adsorption-assisted photocatalysis [20].

Cadmium sulfide (CdS) is an inorganic compound with a narrow bandgap semiconductor material that absorbs most of the visible light in the spectrum. Its appropriate band edge position provides a good visible light absorption ability and excellent photoelectric conversion characteristics, making it widely used in ion detection, sensors, photocatalysis, and other fields [21–29]. In terms of optical properties, CdS still has room for exploration. For example, Yoichi Kobayashi et al. [30] used femtosecond transient absorption spectroscopy to investigate the effects of surface defect states and surface capping reagents on Auger recombination in CdS quantum dots (QDs). They claim that Auger recombination in CdS QDs does not depend on interfacial electronic structures originating from the surface defects and capping reagents of one monolayer level. In addition, Saptarshi Chakraborty et al. [31] used extended X-ray absorption fine structure spectroscopy to study dopant-induced structural perturbations and femtosecond transient absorption (TA) spectroscopy to study ultrafast charge-carrier dynamics, demonstrating that the ionic radius and the dopant oxidation state play a crucial role in determining the dopant anion bond lengths.

CdS also has disadvantages to a certain extent. Its photogenerated carriers are prone to recombination and susceptible to photocorrosion, significantly restricting their widespread application [32]. On the other hand, constructing heterojunctions can effectively promote charge separation, improve photocatalytic performance, and enhance the stability of photocatalytic materials [33]. Therefore, how to modify CdS by incorporating other non-precious metals or biomass charcoal is a compelling research direction. For example, by controlling the oxidation and reduction sites of the CdS/CdSe heterostructure, Rajesh Bera et al. [34] achieved a maximum H_2 generation of 5125 μmol/g/h for a 27.5 wt% CdSe-loaded CdS heterostructure, which was found to be 44 times higher than that of bare CdS nanorods and 22 times higher than that of CdSe nanoparticles.

Moreover, it is important to improve the photostability of CdS. A number of methods have been used to solve this problem, including metal ion doping [35–37], compounding with other semiconductors [38,39], and introducing co-catalysts [40,41]. In recent years, scientists have introduced the use of carbon-based materials such as biochar. Due to their high specific surface area and sufficient surface functional groups, they are considered one of the most effective methods to improve the performance of CdS in photocatalytic hydrogen production [42,43]. For example, the nanocomposites of mesoporous carbon and CdS nanoparticles have been shown to improve photostability and minimize photocorrosion of CdS. The introduction of mesoporous carbon prevents the recombination of photogenerated electrons and holes, thus avoiding photocorrosion in CdS and improving the stability and photocurrent response of the material [44]. Shi et al. [45] prepared a novel biochar-loaded CdS/TiO_2 photocatalyst (CdS/TiO_2/BC) for glucose photoreforming using a simple hydrothermal and calcination method. The as-synthesized CdS/TiO_2/BC exhibited excellent acetic acid selectivity (63.94%) together with improved H_2 generation (~12.77 mmol·g^{-1}·h^{-1}) in 25 mM NaOH solution, while efficient formic acid selectivity (60.29%) and H_2 generation (~10.29 mmol·g^{-1}·h^{-1}) were observed in 3 mM Na_2CO_3 solution.

In this work, a biochar-supported cadmium sulfide (CdS-C) composite photocatalyst is prepared by a hydrothermal method combined with calcination using biochar as a support. A series of characterization analyses are carried out on the prepared samples and the visible light hydrogen production activity is evaluated. The optical and photoelectric properties of as-prepared photocatalysts are also characterized here. The aim of this work is to provide a facile pathway for constructing CdS photocatalytic systems for photocatalytic water decomposition for hydrogen production.

2. Results and Discussion

2.1. Synthesis and Characterizations

CdS–biochar-based composites are usually synthesized by a combined hydrothermal and calcination method [46–52]. In contrast, in this work, orange peel biochar-supported cadmium sulfide composites (CdS-C) were prepared by a combined hydrothermal and calcination method without adding any activation agent.

The XRD patterns of pure CdS and the CdS-xC (x = 20, 40, 60, 80) series of composites are shown in Figure 1, where x represents the amount of biomass char. It can be seen that the peak positions are similar across all samples. Peaks at 26.44°, 43.74°, and 51.82° correspond to the 75-1546 cubic phase, while peaks at 24.87° and 28.09° correspond to the 77-2306 hexagonal phase of the mixed crystalline form [53–55]. After incorporating orange peel biochar into pure CdS, the peak positions remained largely unchanged, though the intensity varied slightly. This indicates that while the crystal structure of the CdS remained unaltered, the crystallinity was affected to varying degrees by the addition of biochar.

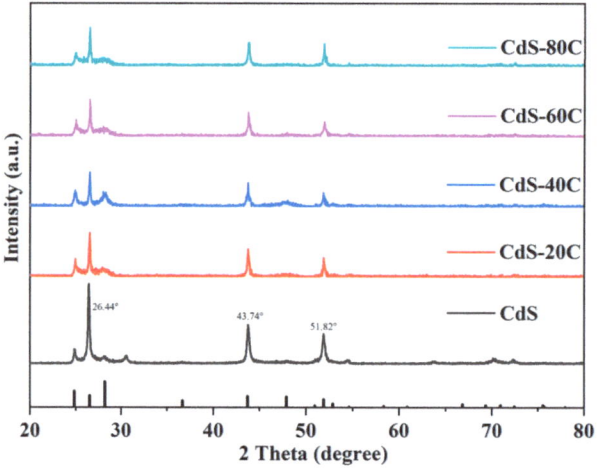

Figure 1. XRD patterns of CdS and CdS-xC (x = 20, 40, 60, 80).

The specific surface area of photocatalytic materials is crucial as it influences their physical adsorption properties [56]. Figure 2a shows the adsorption–desorption isotherm of N_2 on the CdS material, and Figure 2b shows the adsorption–desorption isotherm of N_2 on the CdS-60C material. The addition of orange peel biochar caused slight modifications on the surface of the CdS-C composite materials. This may be attributed to the alkaline etching treatment, which increased the pore density of the biochar, thereby enhancing the specific surface area of the composite material. This increase provides more attachment sites for CdS and enhances the reactivity in photocatalytic hydrogen production. The specific surface area and pore structure data for CdS and the CdS-xC materials are provided in the Supporting Information (Table S1), with CdS-60C exhibiting the largest specific surface area of 6.96 m^2/g. The pore size distribution of CdS-60C is illustrated in Supplementary Figure S1. The reason may be that alkaline etching treatment increased the pores of the orange peel biochar itself, increased the specific surface area of the material, provided more attachment points for CdS, and improved the reactivity of photocatalytic hydrogen production. The specific surface area of CdS and CdS-xC and the pore structure data of the materials are shown in Supplementary Table S1. The CdS-60C material had the relatively largest specific surface area of 6.96 m^2/g. The pore size distribution map of CdS-60C is shown in Supplementary Figure S1.

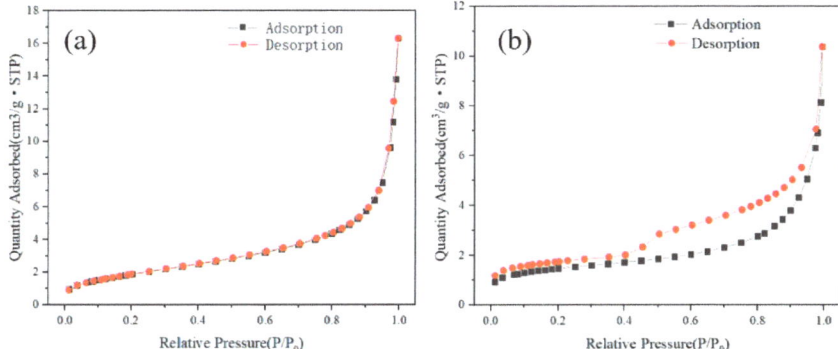

Figure 2. N$_2$ adsorption–desorption isotherms of CdS (**a**) and N$_2$ adsorption–desorption isotherms of CdS-60C (**b**).

The UV–visible diffuse reflectance spectral images of CdS and 20%, 40%, 60%, and 80% orange peel biomass char-loaded CdS materials are shown in Figure 3a. The spectra show slight differences in absorption characteristics among the various composites. Pure CdS exhibits a strong absorption band edge at about 550 nm, which is related to the intrinsic band gap absorption. Furthermore, Figure 3b shows the band gaps (Eg) of the CdS-60C composite and pure CdS, which are about 1.95 eV and 2.21 eV, respectively [57]. In the CdS-xC system, increasing the biochar content from orange peel narrowed the composite material's band gap, resulting in stronger visible light absorption compared to pure CdS, with the absorption peak shifting towards the red and becoming more pronounced. This redshift and enhancement in visible light absorption are primarily due to the incorporation of orange peel biochar, which alters the material's appearance by darkening its color, thereby enhancing its visible light absorption efficiency. Generally, as the biochar ratio increases, the material's visible light absorption band widens, and the band edge experiences a redshift. This suggests that biochar is an effective support for CdS, significantly enhancing visible light absorption, reducing the band gap, and markedly improving the photocatalytic hydrogen production performance. Orange peel biochar shows a broader visible light absorption band, as shown in Figure S2 in the Supporting Information.

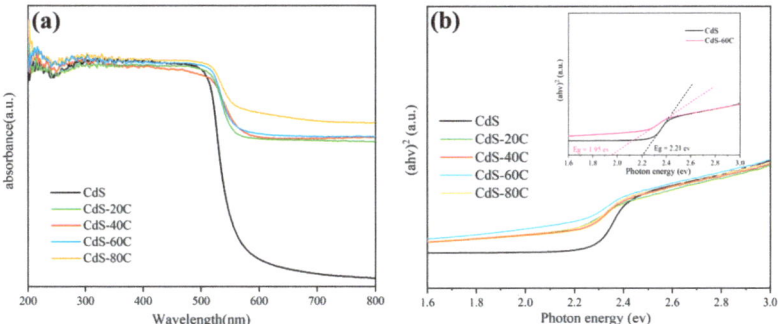

Figure 3. UV–Vis DRS spectra (**a**) and K-M plots (**b**) of CdS and CdS-xC composite materials.

To further analyze the materials, the CdS-60C composite was observed by transmission electron microscopy (TEM). Figure 4 presents TEM images of CdS-60C. It is obvious that the orange peel biochar exhibits a multi-layered sheet structure with many small pores on its surface, which provides a very good site for the growth of CdS (Figure 4a,b). The CdS is distributed on the orange peel biochar, and they show very good binding properties (Figure 4c,d). The crystal structure characteristics of the CdS-60C sample were further examined using high-resolution TEM (HRTEM) (Figure 4f–h). The lattice spacings of the CdS-60C material are 0.335 nm and 0.356 nm, corresponding to the (002) and (100) crystal planes of cubic phase CdS, respectively. Moreover, the uniform distribution of CdS nanoparticles on the orange peel biochar can reduce the aggregation of nanoparticles and provide more reactive sites, which can enhance the photocatalytic performance.

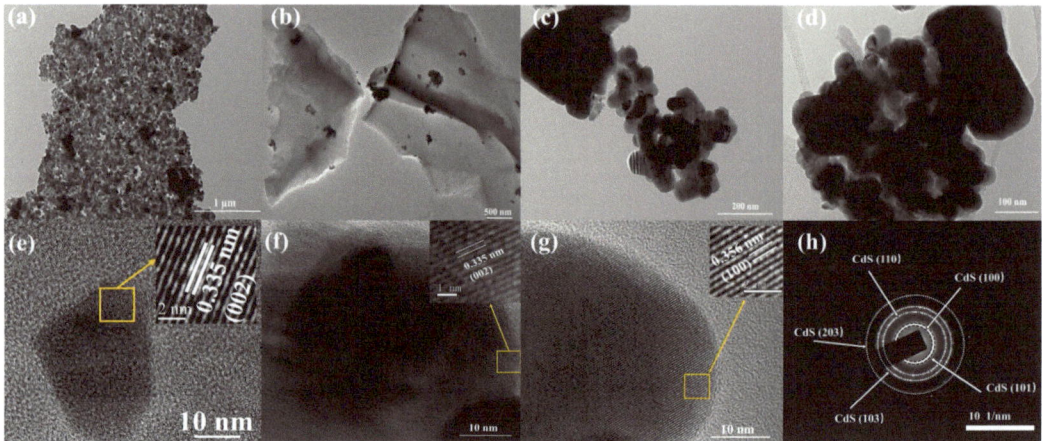

Figure 4. TEM (**a**–**d**), HRTEM (**e**–**g**), and the corresponding SAED (**h**) images of CdS-60C.

In order to explore the mechanism of the improved efficiency of CdS-60C's photocatalytic hydrogen production, photoluminescence spectroscopy was used to investigate photogenerated electron–hole generation and transport in the composites. As shown in Figure 5, pure CdS exhibits a broad green emission band and a strong red emission band, with an excitation wavelength of 450 nm. However, the luminescence intensity of CdS-60C decreases sharply compared with pure CdS, which suggests that the successful interfacial contact can effectively transport the photogenerated carriers and hinder the recombination of electron–hole pairs in CdS.

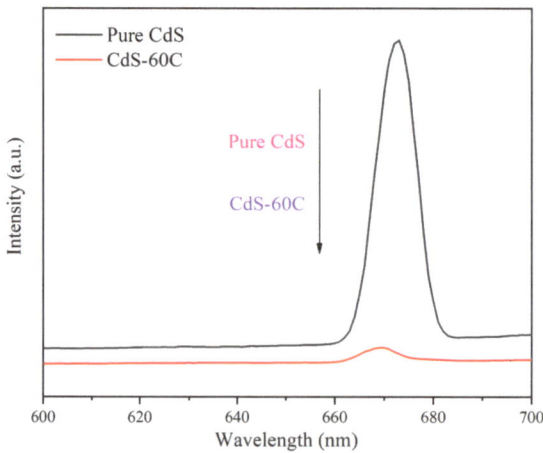

Figure 5. PL spectra of pure CdS and CdS-60C composite.

The ability to achieve efficient photogenerated electron–hole separation and transfer is a very critical factor in improving the photocatalytic hydrogen production activity of a material. The speed of separation and transfer of photoelectron–hole pairs of a material can be obtained from the side by measuring the transient photocurrent of the material. The transient photocurrent response curves of CdS and CdS-60C are shown in Figure 6. It can be seen from the image that the photocurrent of the CdS-60C composite material increases significantly compared with pure CdS. Obviously, biochar acts as a temporary receiver of photogenerated electrons and can quickly absorb a large number of photogenerated electrons generated on the CdS surface. At the same time, biochar can also effectively inhibit the rapid recombination of photogenerated electron–hole pairs, so that under the same external conditions, CdS-60C has a stronger photocurrent; thus, the material has better photocatalytic activity.

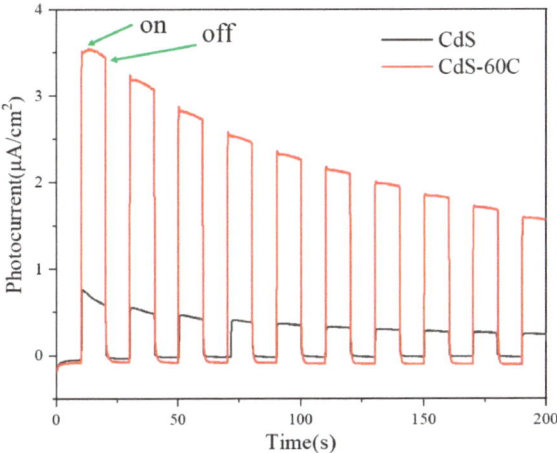

Figure 6. Transient photocurrent response test images for CdS and CdS-60C.

Electrochemical impedance spectroscopy (EIS) was used to further confirm that CdS-60C has a faster electron transfer rate. As shown in Figure 7, from the curve we can see that the curvature radius of the CdS-60C material is lower than that of pure CdS, which indicates that the resistance of the CdS-60C material is lower. These results fully demonstrate the advantages of orange peel biochar in charge loading transfer and recombination delay and explain the improved photocatalytic activity of CdS-60C.

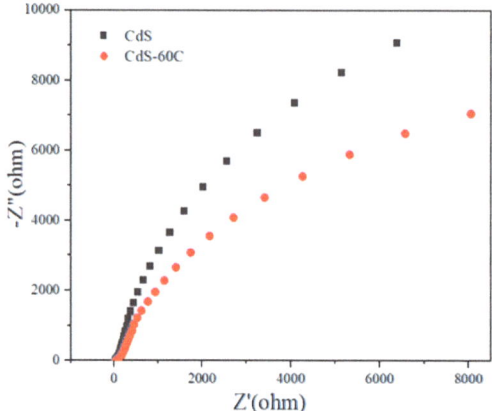

Figure 7. Electrochemical AC impedance of CdS and CdS-60C.

2.2. Hydrogen Production Testing of Materials

Under equivalent sunlight irradiation, the production distribution diagram of photocatalytic hydrogen production by CdS and CdS-xC in each period is illustrated in Figure 8. Each material in the system underwent photocatalysis to produce hydrogen for 4 h under identical lighting conditions. Among them, the hydrogen production rate of pure CdS was 2.11 mmol·g^{-1}·h^{-1}. Notably, the material with the highest hydrogen rate in this system was CdS-60C, achieving 7.8 mmol·g^{-1}·h^{-1} under visible light conditions, which is 3.69 times higher than the pure CdS hydrogen production rate. Furthermore, compared with pure CdS, the hydrogen production rates of the composite materials with varying proportions of biochar in the system increased to different degrees. Specifically, their hydrogen production rates were 2.8 mmol·g^{-1}·h^{-1}, 4.3 mmol·g^{-1}·h^{-1}, 7.8 mmol·g^{-1}·h^{-1}, and 4.1 mmol·g^{-1}·h^{-1}, respectively, representing increases of 1.33, 2.04, 3.69, and 1.94 times compared with the blank control group. Besides, the comparison of the photocatalytic H2 production performance of other het-erojunction and CdS reported previously in the literature is shown in Table S3 [58–67], demonstrating that cDS-60C has the advantage in pure water systems.

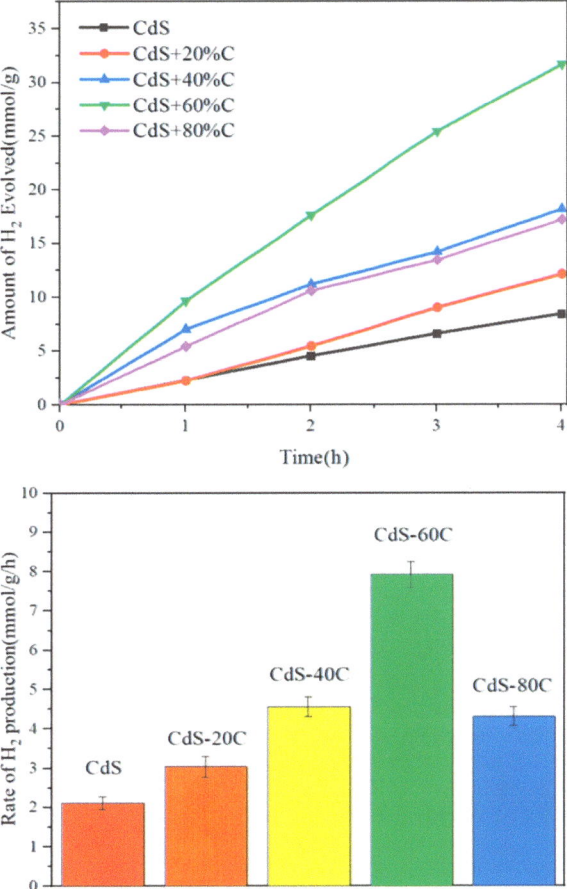

Figure 8. Visible light catalytic hydrogen production efficiency diagram of CdS and CdS-xC (x = 20, 40, 60, 80).

2.3. Cyclic Stability Testing of Materials for Photocatalytic Hydrogen Production

To assess the stability of the photocatalyst, a photocatalytic hydrogen production cycle test was conducted on CdS-60C. As depicted in Figure 9a, after three consecutive cycles of photocatalytic hydrogen production over 12 h, the hydrogen production rate of CdS-60C remained at 25.2 mmol·g^{-1}·h^{-1}, representing 83% of the initial hydrogen production rate. It is worth noting that due to the consumption of sacrificial agents in each cycle, the rate of H$_2$ release tended to decrease. However, upon replacing the sacrificial agent, the photocatalytic hydrogen evolution activity of the CdS-60C composite was restored, indicating its good photostability. After the cycling experiments, the XRD patterns of the regenerated samples were almost the same as those of the fresh ones. As shown in Figure 9b, after three cycles, the crystal structure of CdS did not change, but the intensity of some characteristic peaks weakened, which indicated that the prepared photocatalysts had good stability and recyclability. The slight decrease in photocatalytic efficiency may be caused by partial loss of the photocatalyst during washing.

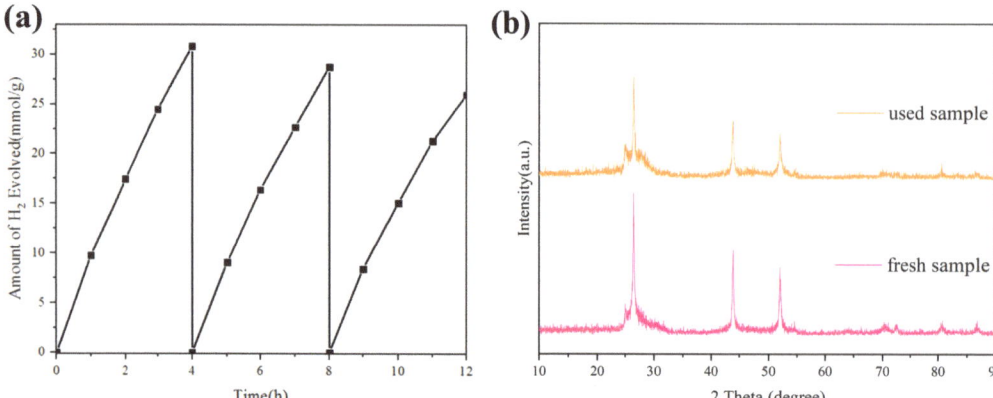

Figure 9. Cycle stability experiment diagram of CdS-60C (**a**) and XRD pattern of CdS-60C composite after three cyclic tests (**b**).

2.4. Reaction Mechanism of Photocatalytic Decomposition of Water for Hydrogen Production

Through the summary of the above experiments and characterizations, it becomes evident that the addition of orange peel biochar not only altered the color appearance of CdS but also influenced its photocatalytic hydrogen production rate and cycle stability. The potential mechanism of the photocatalytic hydrogen evolution reaction of the CdS-60C composite is depicted in Figure 10. Upon exposure to light, CdS is excited by visible light, generating electrons and holes. These electrons can transfer from CdS to biochar, where hydrogen ions in the water can accept the electrons and undergo reduction to release hydrogen, thus enhancing the photocatalytic hydrogen evolution performance. The results of the transient photocurrent response, PL spectra, and EIS experiments further support these findings. Additionally, biochar exhibits a strong photothermal effect due to its efficient absorption of near-infrared light, thereby kinetically promoting the photocatalytic hydrogen production reaction by increasing the local temperature. Moreover, under simulated sunlight irradiation, the principal reactions of CdS-60C heterojunction photocatalytic decomposition of aqueous hydrogen in solution are as follows:

$$CdS/C + h\nu \rightarrow CdS(e_{CB}^- + h_{VB}^+)/C(e_{CB}^- + h_{VB}^+) \quad (1)$$

$$CdS(h_{VB}^+) + C(e_{CB}^-) \rightarrow CdS/C(e_{CB}^- + h_{VB}^+) \quad (2)$$

$$2H^+ + 2e_{CB}^- \rightarrow H_2 \uparrow \quad (3)$$

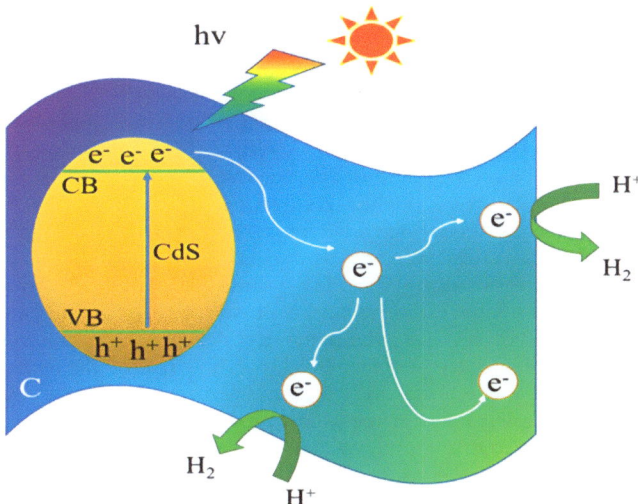

Figure 10. Photocatalytic hydrogen production mechanism of CdS-60C.

3. Experimental

3.1. Preparation of Biochar Material from Orange Peels

The schematic diagram of the preparation of the orange peel biocarbon material is shown in Figure 11. Firstly, citrus peels were collected and dried thoroughly in the sunlight on a sunny day. The dried orange peels were thoroughly crushed with a grinder to produce an orange-colored fine powder after processing. Secondly, the powder made from the obtained batch of orange peels was evenly spread into a quartz boat. The quartz boat containing the orange peel powder was placed in a programmable heating tube furnace, with the temperature set at 500 °C, a heating rate of 5 °C/min, and a constant temperature time of two hours at 500 °C. Finally, according to the program, it naturally cooled to room temperature. During the heating period, high-purity nitrogen was continuously passed into the tube furnace as a protective carrier gas, which was used to avoid the occurrence of combustion or even explosion of the orange peel powder when it met with oxygen at high temperatures.

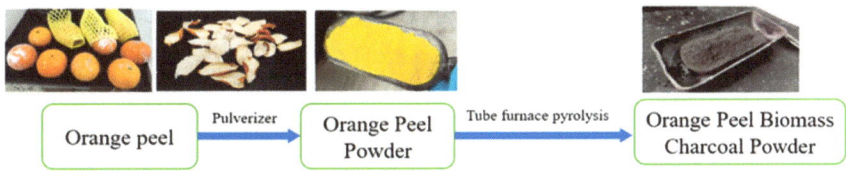

Figure 11. Schematic diagram of the preparation of orange peel biocarbon materials.

The next process was alkaline etching of the orange peel biomass charcoal. The first step was to prepare a potassium hydroxide solution. Here, 11.222 g of potassium hydroxide solid (0.2 mol) was weighed and dissolved in 200 mL of deionized water to obtain a dilute solution of potassium hydroxide (KOH) at a concentration of 1 mol/L. Secondly, the orange peel biochar made in the previous step was sufficiently ground, and 8 g was taken and poured into the solution, while the speed of the magnetometer stirrer was adjusted to 360 r/min for constant stirring of the mixture for 24 h. The purpose of this step was to carry out sufficient alkali etching of the orange peel biomass charcoal, such that a lot of fine pores could be etched on the surface of the orange peel biochar. After the etching was completed, the KOH solution was poured off and the orange peel biomass char was rinsed with deionized water until the soaking solution was neutral, and then the cleaned orange peel biomass char was freeze-dried. After freeze-drying, the orange peel biomass charcoal was ground again, and the orange peel biochar obtained after sufficient grinding was collected for use in the subsequent steps.

3.2. Preparation of CdS Blank Control Material

First, 70 mL of deionized water, 3 mmol of $Cd(CH_3COO)_2\text{-}2H_2O$, and 4 mmol of thiourea were weighed. A stirrer and the weighed deionized water, $Cd(CH_3COO)_2\text{-}2H_2O$, and thiourea were added into a 200 mL beaker; the magnetic stirrer was turned on with the rotational speed adjusted to 450 r/min, and the stirring process lasted 6 h. The stirring process was carried out in the same order. After the stirring was completed, the solution was poured into a reactor liner made of polytetrafluoroethylene (PTFE), the reactor liner was covered and put into the steel sleeve of the reactor, and the lid of the steel sleeve was tightened to ensure that it had been fully sealed. The reaction kettle was placed in the oven at a temperature of 170 °C, and the reaction time was set to 20 h. After the reaction was completed, the orange-yellow product obtained in the reaction kettle was transferred to a centrifuge tube and washed at least three times with 99.5% ethanol and ultrapure water; then, the product was placed in an oven, and the drying temperature was set at 60 °C for 10 h. After the materials were thoroughly dried, the materials were fully ground and the the powdered samples were collected, sealed, and stored away from light for subsequent photocatalytic hydrogen production performance testing and characterization analysis.

3.3. Preparation of CdS-C Composites

The weighing and operation of the sulfur and cadmium sources were the same as in Section 3.2. In order to obtain the theoretical content of CdS in the composite, i.e., 20%, 40%, 60%, and 80%, calculated based on the theoretical yield, the corresponding amount of biochar used was 57.79 mg, 115.48 mg, 173.37 mg, and 231.16 mg, respectively. The weighed biochar was added to the beaker along with the sulfur source and cadmium source and stirred. After the stirring was completed, the solution was transferred to a reaction kettle, sealed, and placed in an oven with the temperature set at 170 °C and the reaction time at 20 h. The obtained samples based on the theoretical content of CdS in the composite were marked as CdS-20C, CdS-40C, CdS-60C, and CdS-80C, respectively.

3.4. Characterizations

The crystal structure was analyzed using X-ray diffraction (XRD) (Rigaku Co., Tokyo, Japan) with Cu Kα radiation at a running voltage of 36 kV and a 20 mA current. The sample morphology, elemental mapping, and lattice fringes were examined using transmission electron microscopy (TEM) (JEM-2100, Tokyo, Japan). UV–visible diffuse reflectance spectroscopy (UV-2401PC, Shimadzu, Tokyo, Japan) was used to record the light absorption spectra. The Brunauer–Emmett–Teller (BET) (Micromeritics, Norcross, GA, USA) specific surface area of the samples was determined using a Micromeritics ASAP 2460 nitrogen adsorption–desorption machine. The photoelectrochemical measurements, including photocurrent response (i-t) and electrochemical impedance (EIS), were taken using an electrochemical workstation (CHI-760D, Huake Putian Technology, Beijing, China). The photoluminescence (PL) spectra were recorded on an F97 Pro spectrophotometer (Lengguang Tech., Shanghai, China).

3.5. Photocatalytic Hydrogen Evolution

The photocatalytic hydrogen production activity through water splitting was assessed in a professional system (LabSolar-IIAG, Perfectlight, Beijing, China). In the typical progress, 0.1 mol Na_2SO_3 and 0.05 mol $Na_2S.9H_2O$ were dissolved in a reactor filled with 100 mL ultrapure water to form a homogeneous solution. Then, 20 mg of photocatalyst was uniformly dispersed in the solution and with stirring maintained. Before the measurement, the reactor needed to be degassed for half an hour to ensure that the pressure was less than

−0.12 MPa. During the photocatalytic hydrogen evolution reaction, the suspension was carried out under a 300 Wxenon lamp (CEL-HFX300, Aulight, Beijing, China) with a UV-IR cut-420 filter and cooled to 10 °C with condensed water. Finally, an online GC-9750 gas chromatograph (GC, China Fuli, nitrogen as carrier gas and 5A molecular sieve column) was used to collect and detect the amount of H_2 precipitated at 1-h intervals.

4. Conclusions

In summary, orange peel biochar-supported cadmium sulfide composites (CdS-C) with prominent light absorption performance were prepared using a combined hydrothermal and calcination method. Orange peel biochar incorporation not only increased the photocatalytic H_2 production rates but also greatly reduced the agglomeration of the material itself and improved the cycling stability. The large surface area of the biochar also provided ample space for reactions to occur. It was found that the hydrogen production efficiency of the CdS-C series materials was significantly improved compared to that of CdS without biochar. In particular, CdS-C with 60% CdS in the composite exhibited the highest photocatalytic activity at 7.8 mmol·g^{-1}·h^{-1}, which is about 3.69 times that of pure CdS. It was still stable after three consecutive cycles, indicating that the biochar improved the photostability of the CdS. Semiconductor-loaded biocarbon-based structures have excellent photocatalytic hydrogen production efficiency, paving the way for important potential applications in visible light hydrogen evolution.

Supplementary Materials: The following supporting information can be downloaded at: https://www.mdpi.com/article/10.3390/inorganics12060156/s1, Figure S1. Aperture distribution of CdS-60C; Figure S2. UV-vis DRS spectra of the orange peels biochar. Table S1. Specific surface area and pore structure data of CdS and CdS-xC (x = 20, 40, 60, 80). Table S2. Hydrogen production rates of CdS and CdS(x = 20,40,60,80) normalized by BET surface. Table S3. Comparison of photocatalytic H2 production with other photocatalysts.

Author Contributions: Conceptualization, X.L. and J.W.; methodology, X.L. and M.H.; experimental, Y.Z.; validation, M.H.; investigation, J.Z. (Jindi Zhang); resources, L.Z. and J.W.; data curation, Y.Z.; writing—original draft preparation, X.L.; writing—review and editing, J.W.; visualization, J.Z. (Jing Zhang); supervision, L.Z. and J.W.; funding acquisition, J.W. All authors have read and agreed to the published version of the manuscript.

Funding: This research received no external funding.

Data Availability Statement: The original contributions presented in the study are included in the article, further inquiries can be directed to the corresponding author.

Acknowledgments: This work was supported by a project from the Department of Ecology and Environment of Yunnan Province (202305AM340008), an R&D Project (2022 No. 4) from the Water Resources Department of Yunnan Province, the Key Laboratory of Advanced Materials for Wastewater Treatment of Kunming (2110304), and the Institute of Frontier Technologies in Water Treatment Co., Ltd., Kunming, 650503, China.

Conflicts of Interest: Authors Jiaqiang Wang and Xiang Li are employed by the company, Institute of Frontier Technologies in Water Treatment Co., Ltd. The remaining authors declare that the research was conducted in the absence of any commercial or financial relationships that could be construed as a potential conflict of interest. The Institute of Frontier Technologies in Water Treatment Co., Ltd. was not involved in the study design, collection, analysis, interpretation of data, the writing of this article or the decision to submit it for publication.

References

1. Yamauchi, M.; Saito, H.; Sugimoto, T.; Mori, S.; Saito, S. Sustainable organic synthesis promoted on titanium dioxide using coordinated water and renewable energies/resources. *Coord. Chem. Rev.* **2022**, *472*, 214773–214802. [CrossRef]
2. Liao, G.; Gong, Y.; Zhang, L.; Gao, H.; Yang, G.; Fang, B. Semiconductor polymeric graphitic carbon nitride photocatalysts: The "holy grail" for the photocatalytic hydrogen evolution reaction under visible light. *Energy Environ. Sci.* **2019**, *12*, 2080–2147. [CrossRef]

3. Zhao, Z.; Ge, G.; Zhang, D.; Cheng, B.; Yu, J. Heteroatom-Doped Carbonaceous Photocatalysts for Solar Fuel Production and Environmental Remediation. *ChemCatChem* **2018**, *10*, 62–123. [CrossRef]
4. Kuang, P.; Sayed, M.; Fan, J.; Cheng, B.; Yu, J. 3D Graphene-Based H_2-Production Photocatalyst and Electrocatalyst. *Adv. Energy Mater.* **2020**, *10*, 1–53. [CrossRef]
5. Tang, S.; Xia, Y.; Fan, J.; Cheng, B.; Yu, J.; Ho, W. Enhanced photocatalytic H_2 production performance of CdS hollow spheres using C and Pt as bi-cocatalysts. *J. Catal.* **2021**, *42*, 743–752. [CrossRef]
6. Zhang, Y.; Su, C.; Chen, J.; Huang, W.; Lou, R. Recent progress of transition metal-based biomass-derived carbon composites for supercapacitor. *J. Rare Met.* **2023**, *42*, 769–796. [CrossRef]
7. Jiang, J.; Zhang, Q.; Zhan, X.; Chen, F. Renewable, Biomass-Derived, Honeycomblike Aerogel as a Robust Oil Absorbent with Two-Way Reusability. *ACS Sustain. Chem. Eng.* **2017**, *5*, 10307–10316. [CrossRef]
8. Fang, X.; Li, W.; Chen, X.; Wu, Z.; Zhang, Z.; Zou, Y. Controlling the microstructure of biomass-derived porous carbon to assemble structural absorber for broadening bandwidth. *Carbon* **2022**, *198*, 70–79. [CrossRef]
9. Yao, P.; Zhong, W.; Zhang, Z.; Yang, S.; Gong, Z.; Jia, C.; Chen, P.; Cheng, J.; Chen, Y. Surface engineering of biomass-derived carbon material for efficient water softening. *Chem. Eng. Sci.* **2023**, *282*, 119312–119319. [CrossRef]
10. Chen, G.; Wang, H.; Han, L.; Yang, N.; Hu, B.; Qiu, M.; Zhong, X. Highly efficient removal of U(VI) by a novel biochar supported with FeS nanoparticles and chitosan composites. *J. Mol. Liq.* **2020**, *15*, 45–53. [CrossRef]
11. Jin, C.; Sun, J.; Chen, Y.; Guo, Y.; Han, D.; Wang, R.; Zhao, C. Sawdust wastes-derived porous carbons for CO_2 adsorption. Part 1. Optimization preparation via orthogonal experiment. *Sep. Purif. Technol.* **2021**, *276*, 119270–119281. [CrossRef]
12. Wang, Z.; Shen, D.; Wu, C.; Gu, S. State-of-the-art on the production and application of carbon nanomaterials from biomass. *Green Chem.* **2018**, *20*, 5031–5057. [CrossRef]
13. Jin, C.; Sun, J.; Bai, S.; Zhou, Z.; Sun, Y.; Guo, Y.; Wang, R.; Zhao, C. Sawdust wastes-derived porous carbons for CO_2 adsorption. Part 2. Insight into the CO_2 adsorption enhancement mechanism of low-doping of microalgae. *J. Environ. Chem. Eng.* **2022**, *10*, 108265–108276. [CrossRef]
14. Wang, L.; Zhang, H.; Wang, Y.; Qian, C.; Dong, Q.; Deng, C.; Jiang, D.; Shu, M.; Pan, S.; Zhang, S. Unleashing ultra-fast sodium ion storage mechanisms in interface-engineered monolayer MoS_2/C interoverlapped superstructure with robust charge transfer networks. *J. Mater. Chem. A* **2020**, *8*, 15002–15011. [CrossRef]
15. Lu, Y.; Cai, Y.; Zhang, S.; Zhuang, L.; Hu, B.; Wang, S.; Chen, J.; Wang, X. Application of biochar-based photocatalysts for adsorption-(photo)degradation/reduction of environmental contaminants: Mechanism, challenges and perspective. *Biochar* **2022**, *4*, 45–69. [CrossRef]
16. Tan, X.; Liu, Y.; Gu, Y.; Xu, Y.; Zeng, G.; Hu, X.; Liu, S.; Wang, X.; Liu, S.; Li, J. Biochar-based nano-composites for the decontamination of wastewater: A review. *Bioresour. Technol.* **2016**, *212*, 318–333. [CrossRef] [PubMed]
17. Lee, J.; Park, Y.-K. Applications of modified biochar-based materials for the removal of environment pollutants: A mini review. *Sustainability* **2020**, *12*, 6112–6127. [CrossRef]
18. Abhijeet, P.; Prem, P.; Chen, X.; Balasubramanian, P.; Chang, S. Activation methods increase biochar's potential for heavy-metal adsorption and environmental remediation: A global meta-analysis. *Sci. Total Environ.* **2023**, *865*, 161252–161262.
19. Talukdar, K.; Jun, B.; Yoon, Y.; Kim, Y.; Fayyaz, A.; Park, C. Novel Z-scheme Ag_3PO_4/Fe_3O_4-activated biochar photocatalyst with enhanced visible-light catalytic performance toward degradation of bisphenol A. *J. Hazard. Mater.* **2020**, *398*, 123025–123034. [CrossRef]
20. Kumar, A.; Kumar, A.; Sharma, G.; Naushad, M.; Stadler, F.; Ghfar, A.; Dhiman, P.; Saini, R. Sustainable nano-hybrids of magnetic biochar supported g-C_3N_4/$FeVO_4$ for solar powered degradation of noxious pollutants- Synergism of adsorption, photocatalysis & photo-ozonation. *J. Clean. Prod.* **2017**, *165*, 431–451. [CrossRef]
21. He, J.; Chen, L.; Wang, F.; Liu, Y.; Chen, P.; Au, C.; Yin, S. CdS Nanowires Decorated with Ultrathin MoS_2 Nanosheets as an Efficient Photocatalyst for Hydrogen Evolution. *ChemSusChem* **2016**, *9*, 624–630. [CrossRef] [PubMed]
22. Hu, Y.; Gao, X.; Yu, L.; Wang, Y.; Ning, J.; Xu, S.; Wen, X.; Lou, D. Carbon-Coated CdS Petalous Nanostructures with Enhanced Photostability and Photocatalytic Activity. *Angew. Chem.* **2013**, *125*, 5746–5749. [CrossRef]
23. Kim, M.; Kim, Y.; Lim, S.; Kim, S.; In, S. Efficient visible light-induced H_2 production by Au@CdS/TiO_2 nanofibers: Synergistic effect of core–shell structured Au@CdS and densely packed TiO_2 nanoparticles. *Appl. Catal. B-Environ.* **2015**, *16*, 423–431. [CrossRef]
24. Li, X.; Dong, H.; Wang, B.; Lv, J.; Xu, G.; Wang, D.; Wu, Y. Controllable Synthesis of MoS_2/h-CdS/c-CdS Nanocomposites with Enhanced Photocatalytic Hydrogen Evolution Under Visible Light Irradiation. *Catal. Lett.* **2018**, *148*, 67–70. [CrossRef]
25. Wang, X.; Liu, G.; Wang, L.; Chen, Z.; Lu, G.; Cheng, H. ZnO–CdS@Cd Heterostructure for Effective Photocatalytic Hydrogen Generation. *Adv. Energy Mater.* **2012**, *16*, 81–83. [CrossRef]
26. Yu, X.; Du, R.; Li, B.; Zhang, Y.; Liu, H.; Qu, J.; An, X. Biomolecule-assisted self-assembly of CdS/MoS_2/graphene hollow spheres as high-efficiency photocatalysts for hydrogen evolution without noble metals. *Appl. Catal. B* **2016**, *3*, 51–55.
27. Ye, A.; Fan, W.; Zhang, Q.; Deng, W.; Wang, Y. CdS–graphene and CdS–CNT nanocomposites as visible-light photocatalysts for hydrogen evolution and organic dye degradation. *Catal. Sci. Technol.* **2012**, *2*, 969–978. [CrossRef]
28. Zhu, C.; Liu, C.; Fu, Y.; Gao, J.; Huang, H.; Liu, Y.; Kang, Z. Construction of CDs/CdS photocatalysts for stable and efficient hydrogen production in water and seawater. *Appl. Catal. B* **2019**, *242*, 78–85. [CrossRef]

29. Schneider, J.; Bahnemann, D. Undesired Role of Sacrificial Reagents in Photocatalysis. *J. Phys. Chem. Lett.* **2013**, *4*, 3479–3483. [CrossRef]
30. Kobayashi, Y.; Nishimura, T.; Yamaguchi, H.; Tamai, N. Effect of Surface Defects on Auger Recombination in Colloidal CdS Quantum Dots. *J. Phys. Chem. Lett.* **2011**, *2*, 1051–1055. [CrossRef]
31. Chakraborty, S.; Mondal, P.; Makkar, M.; Moretti, L.; Cerullo, G.; Viswanatha, R. Transition Metal Doping in CdS Quantum Dots: Diffusion, Magnetism, and Ultrafast Charge Carrier Dynamics. *Chem. Mater.* **2023**, *35*, 2146–2154. [CrossRef]
32. Wang, P.; Xu, S.; Wang, J.; Liu, X. Photodeposition synthesis of CdS QDs-decorated TiO_2 for efficient photocatalytic degradation of metronidazole under visible light. *J. Mater. Sci. Mater. Electron.* **2020**, *31*, 19797–19808. [CrossRef]
33. Feng, Y.; Li, J.; Ye, S.; Gao, S.; Cao, R. Growing COFs in situ on CdS nanorods as core–shell heterojunctions to improve the charge separation efficiency. *Sustain. Energy Fuels* **2022**, *6*, 5089–5099. [CrossRef]
34. Bera, R.; Dutta, A.; Kundu, S.; Polshettiwar, V.; Patra, A. Design of a CdS/CdSe Heterostructure for Efficient H_2 Generation and Photovoltaic Applications. *J. Phys. Chem. C* **2018**, *122*, 12158–12167. [CrossRef]
35. Korake, P.; Achary, S.; Gupta, N. Role of aliovalent cation doping in the activity of nanocrystalline CdS for visible light driven H_2 production from water. *Int. J. Hydrogen Energy* **2015**, *40*, 8695–8705. [CrossRef]
36. Huang, S.; Lin, Y.; Yang, J.; Li, X.; Zhang, J.; Yu, J.; Shi, H.; Wang, W.; Yu, Y. Enhanced photocatalytic activity and stability of semiconductor by Ag doping and simultaneous deposition: The case of CdS. *RSC Adv.* **2013**, *3*, 20782–20792. [CrossRef]
37. Huang, G.; Zhu, Y. Enhanced photocatalytic activity of $ZnWO_4$ catalyst via fluorine doping. *J. Phys. Chem. C* **2007**, *111*, 11952–11958. [CrossRef]
38. Reddy, D.; Park, H.; Ma, R.; Kumar, D.; Lim, M.; Kim, T. Heterostructured WS_2-MoS_2 ultrathin nanosheets integrated on CdS nanorods to promote charge separation and migration and improve solar-driven photocatalytic hydrogen evolution. *ChemSusChem* **2017**, *10*, 1563–1570. [CrossRef] [PubMed]
39. Reddy, D.; Choi, J.; Lee, S.; Kim, Y.; Hong, S.; Kumar, D.; Kim, T. Hierarchical dandelion-flower-like cobalt-phosphide modified CdS/reduced graphene oxide-MoS_2 nanocomposites as a noble-metal-free catalyst for efficient hydrogen evolution from water. *Catal. Sci. Technol.* **2016**, *6*, 6197–6206. [CrossRef]
40. Yuan, Y.; Li, Z.; Wu, S.; Chen, D.; Yang, L.; Cao, D.; Tu, W.; Yu, Z.; Zou, Z. Role of two-dimensional nanointerfaces in enhancing the photocatalytic performance of 2D–2D MoS_2/CdS photocatalysts for H_2 production. *Chem. Eng. J.* **2018**, *350*, 335–343.
41. Li, L.; Wu, J.; Liu, B.; Liu, X.; Li, C.; Gong, Y.; Huang, Y.; Pan, L. NiS sheets modified CdS/reduced graphene oxide composite for efficient visible light photocatalytic hydrogen evolution. *Catal. Today* **2018**, *315*, 110–116. [CrossRef]
42. Zhang, L.; Zhang, C.; Li, J.; Sun, K.; Zhang, J.; Huang, M.; Wang, J. Hydrothermally prepared unactivated bean sprouts biochar supported CdS with significantly enhanced photocatalytic hydrogen evolution activity. *Catal. Commun.* **2024**, *187*, 106861–106869. [CrossRef]
43. Kang, F.; Shi, C.; Li, W.; Eqi, M.; Liu, Z.; Zheng, X.; Huang, Z. Honeycomb like CdS/sulphur-modified biochar composites with enhanced adsorption-photocatalytic capacity for effective removal of rhodamine B. *J. Environ. Chem. Eng.* **2022**, *10*, 106942–106953. [CrossRef]
44. Banerjee, R.; Pal, A.; Ghosh, D.; Ghosh, A.; Nandi, M.; Biswas, P. Improved photocurrent response, photostability and photocatalytic hydrogen generation ability of CdS nanoparticles in presence of mesoporous carbon. *Mater. Res. Bull.* **2021**, *134*, 111085–111093. [CrossRef]
45. Shi, C.; An, Y.; Gao, G.; Xue, J.; Algadi, H.; Huang, Z.; Guo, Z. Insights into Selective Glucose Photoreforming for Coproduction of Hydrogen and Organic Acid over Biochar-Based Heterojunction Photocatalyst Cadmium Sulfide/Titania/Biochar. *ACS Sustain. Chem. Eng.* **2024**, *12*, 2538–2549. [CrossRef]
46. Huang, H.; Wang, Y.; Jiao, W.; Cai, F.; Shen, M.; Zhou, S.; Cao, H.; Lü, J.; Cao, R. Lotus-Leaf-Derived Activated-Carbon-Supported Nano-CdS as Energy-Efficient Photocatalysts under Visible Irradiation. *ACS Sustain. Chem. Eng.* **2018**, *6*, 7871–7879. [CrossRef]
47. Norouzi, O.; Kheradmand, A.; Jiang, Y.; Maria, F.; Masek, O. Superior activity of metal oxide biochar composite in hydrogen evolution under artificial solar irradiation: A promising alternative to conventional metal-based photocatalysts. *Int. J. Hydrogen Energy* **2019**, *44*, 28698–29708. [CrossRef]
48. Chen, D.; Wang, X.; Zhang, X.; Yang, Y.; Xu, Y.; Qian, G. Facile fabrication of mesoporous biochar/$ZnFe_2O_4$ composite with enhanced visible-light photocatalytic hydrogen evolution. *Int. J. Hydrogen Energy* **2019**, *44*, 19967–19977. [CrossRef]
49. Qian, J.; Chen, Z.; Sun, H.; Chen, F.; Xu, X.; Wu, P.; Li, P.; Ge, W. Enhanced photocatalytic H_2 production on three-dimensional porous CeO_2/Carbon nanostructure. *ACS Sustain. Chem. Eng.* **2018**, *8*, 9691–9698. [CrossRef]
50. Zhou, M.; Zhang, K.; Chen, F.; Chen, Z. Synthesis of biomimetic cerium oxide by bean sprouts bio-template and its photocatalytic performance. *J. Rare Earths* **2016**, *34*, 683–688. [CrossRef]
51. Wei, X.; Ou, C.; Fang, S.; Zheng, X.; Zheng, G.; Guan, X. One-pot self-assembly of 3D CdS-graphene aerogels with superior adsorption capacity and photocatalytic activity for water purification. *Powder Technol.* **2019**, *345*, 213–222. [CrossRef]
52. Huang, H.; Wang, Y.; Cai, F.; Jiao, W.; Zhang, N.; Liu, C.; Cao, H.; Lü, J. Photodegradation of Rhodamine B over biomass-derived activated carbon supported CdS nanomaterials under visible irradiation. *Front. Chem.* **2017**, *5*, 123–133. [CrossRef] [PubMed]
53. Lei, Y.; Yang, C.; Hou, J.; Wang, F.; Min, S.; Ma, X.; Jin, Z.; Xu, J.; Lu, G.; Huang, K. Strongly coupled CdS/graphene quantum dots nanohybrids for highly efficient photocatalytic hydrogen evolution: Unraveling the essential roles of graphene quantum dots. *Appl. Catal. B-Environ.* **2017**, *126*, 59–69. [CrossRef]

54. Zou, L.; Wang, H.; Wang, X. High Efficient Photodegradation and Photocatalytic Hydrogen Production of CdS/BiVO$_4$ Heterostructure through Z-Scheme Process. *ACS Sustain. Chem. Eng.* **2016**, *6*, 6–10. [CrossRef]
55. Liu, S.; Ma, Y.; Chi, D.; Sun, Q.; Chen, Q.; Zhang, J.; He, Z.; He, L.; Zhang, K.; Liu, B. Hollow heterostructure CoS/CdS photocatalysts with enhanced charge transfer for photocatalytic hydrogen production from seawater. *Int. J. Hydrogen Energy* **2022**, *47*, 9220–9229. [CrossRef]
56. Bai, J.; Chen, W.; Shen, R.; Jiang, Z.; Zhang, P.; Liu, W.; Li, X. Regulating interfacial morphology and charge-carrier utilization of Ti$_3$C$_2$ modified all-sulfide CdS/ZnIn$_2$S$_4$ S-scheme heterojunctions for effective photocatalytic H$_2$ evolution. *J. Mater. Sci. Technol.* **2022**, *112*, 85–95. [CrossRef]
57. Bai, J.; Shen, R.; Jiang, Z.; Zhang, P.; Li, Y.; Li, X. Integration of 2D layered CdS/WO$_3$ S-scheme heterojunctions and metallic Ti$_3$C$_2$ MXene-based Ohmic junctions for effective photocatalytic H$_2$ generation. *Chin. J. Catal.* **2022**, *43*, 359–369. [CrossRef]
58. Long, H.; Wang, P.; Wang, X.; Chen, F.; Yu, H. Optimizing hydrogen adsorption of Ni$_x$B cocatalyst by integrating P atom for enhanced photocatalytic H$_2$-production activity of CdS. *Appl. Surf. Sci.* **2022**, *604*, 154457. [CrossRef]
59. Lu, H.; Liu, Y.; Zhang, S.; Wan, J.; Wang, X.; Deng, L.; Kan, J.; Wu, G. Clustered tubular S-scheme ZnO/CdS heterojunctions for enhanced photocatalytic hydrogen production. *Mater. Sci. Eng. B* **2023**, *289*, 116282. [CrossRef]
60. He, K. ZnO/ZnS/CdS three-phase composite photocatalyst with a flower cluster structure: Research on its preparation and photocatalytic activity hydrogen production. *Int. J. Hydrogen Energy* **2024**, *51*, 30–40. [CrossRef]
61. Liu, Y.; Dai, F.; Zhao, R.; Huai, X.; Han, J.; Wang, L. Aqueous synthesis of core/shell/shell CdSe/CdS/ZnS quantum dots for photocatalytic hydrogen generation. *J. Mater. Sci.* **2019**, *54*, 8571–8580. [CrossRef]
62. Liu, Z.; Zhuang, Y.; Dong, L.; Mu, H.; Li, D.; Zhang, F.; Xu, H.; Xie, H. Enhancement Mechanism of Photocatalytic Hydrogen ProductionActivity of CeO$_2$/CdS by Morphology Regulation. *ACS Appl. Energy Mater.* **2023**, *6*, 7722–7736. [CrossRef]
63. Qi, Z.; Chen, J.; Li, Q.; Wang, N.; Carabineiro, S.; Lv, K. Increasing the Photocatalytic Hydrogen Generation Activity of CdS Nanorods by Introducing Interfacial and Polarization Electric Fields. *Small* **2023**, *19*, 2303318. [CrossRef]
64. Liu, J.; Qiu, L.; Liu, Z.; Tang, Y.; Cheng, L.; Chen, Z.; Li, P.; Cao, B.; Chen, X.; Kita, H.; et al. Boosting the photocatalytic activity for H2 production of Bi$_2$O$_2$Se/CdS heterojunction. *Mater. Lett.* **2023**, *345*, 134498. [CrossRef]
65. Meng, A.; Zhu, B.; Zhong, B.; Zhang, L.; Cheng, B. Direct Z-scheme TiO$_2$/CdS hierarchical photocatalyst for enhanced photocatalytic H$_2$ production activity. *Appl. Surf. Sci.* **2017**, *422*, 518–527. [CrossRef]
66. Kalia, R.; Pirzada, B.; Kunchala, R.; Naidu, B. Noble metal free efficient photocatalytic hydrogengeneration by TaON/CdS semiconductor nanocomposites under natural sunlight. *Int. J. Hydrogen Energy* **2023**, *48*, 16246–16258. [CrossRef]
67. Zhang, L.; Zhu, X.; Zhao, Y.; Zhang, P.; Chen, J.; Jiang, J.; Xie, T. The photogenerated charge characteristics in Ni@NiO/CdS hybrids for increased photocatalytic H$_2$ generation. *RSC Adv.* **2019**, *9*, 39604–39610. [CrossRef]

Disclaimer/Publisher's Note: The statements, opinions and data contained in all publications are solely those of the individual author(s) and contributor(s) and not of MDPI and/or the editor(s). MDPI and/or the editor(s) disclaim responsibility for any injury to people or property resulting from any ideas, methods, instructions or products referred to in the content.

Article

Lanthanide-Containing Polyoxometalate Crystallized with Bolaamphiphile Surfactants as Inorganic–Organic Hybrid Phosphors

Rieko Ishibashi [1], Ruka Koike [1], Yoriko Suda [2], Tatsuhiro Kojima [3], Toshiyuki Sumi [1], Toshiyuki Misawa [1], Kotaro Kizu [1], Yosuke Okamura [4] and Takeru Ito [1,*]

[1] Department of Chemistry, School of Science, Tokai University, Hiratsuka 259-1292, Japan
[2] Department of Electric and Electronic Engineering, School of Engineering, Tokyo University of Technology, Hachioji 192-0982, Japan
[3] Department of Applied Chemistry, Kobe City College of Technology, Kobe 651-2194, Japan
[4] Department of Applied Chemistry, School of Engineering, Tokai University, Hiratsuka 259-1292, Japan
* Correspondence: takeito@tokai.ac.jp

Abstract: Lanthanide elements such as europium exhibit distinctive emissions due to the transitions of inner-shell 4f electrons. Inorganic materials containing lanthanide elements have been widely used as phosphors in conventional displays. The hybridization of lanthanide ions with organic components enables to control of the material's shapes and properties and broadens the possibility of lanthanide compounds as inorganic–organic materials. Lanthanide ion-containing polyoxometalate anions (Ln-POM) are a promising category as an inorganic component to design and synthesize inorganic–organic hybrids. Several inorganic–organic Ln-POM systems have been reported by hybridizing with cationic surfactants as luminescent materials. However, single-crystalline ordering has not been achieved in most cases. Here, we report syntheses and structures of inorganic–organic hybrid crystals of lanthanide-based POM and bolaamphiphile surfactants with two hydrophilic heads in one molecule. An emissive decatungstoeuropate ($[EuW_{10}O_{36}]^{9-}$, EuW_{10}) anion was employed as a lanthanide source. The bolaamphiphile counterparts are 1,8-octamethylenediammonium ($[H_3N(CH_2)_8NH_3]^{2+}$, C_8N_2) and 1,10-decamethylenediammonium ($[H_3N(CH_2)_{10}NH_3]^{2+}$, $C_{10}N_2$). Both hybrid crystals of C_8N_2-EuW_{10} and $C_{10}N_2$-EuW_{10} were successfully obtained as single crystals, and their crystal structures were unambiguously determined using X-ray diffraction measurements. The photoluminescence properties of C_8N_2-EuW_{10} and $C_{10}N_2$-EuW_{10} were investigated by means of steady-state and time-resolved spectroscopy. The characteristic emission derived from the EuW_{10} anion was retained after the hybridization process.

Keywords: polyoxometalate; surfactant; single crystal; inorganic–organic hybrid; photoluminescence

1. Introduction

Lanthanide elements can attribute several functions to materials, which have been applied to ionic conductors [1], magnetic materials [2], and biological reagents [3]. One of the most distinctive characteristics are emission properties [4]. Some lanthanides such as europium and terbium exhibit distinctive emission due to the transitions of inner-shell 4f electrons. Inorganic lanthanide compounds have been widely employed as phosphors in conventional displays and imaging technologies. The combination of lanthanide ions with organic moiety enables to control of the material's shapes and properties and broadens the application areas of lanthanide compounds as inorganic–organic hybrid luminescent materials [5–9].

As for the lanthanide source, lanthanide-containing polyoxometalate (Ln-POM) anions are promising inorganic components [10–13]. Various types of Ln-POM anions have been synthesized as single crystals under ambient and/or hydrothermal conditions [14–17]. Hybridizing Ln-POM with organic moieties achieves controllable inorganic—organic hybrid materials in their luminescent properties and material shapes [18–21]. Decatungstoeuropate ($[EuW_{10}O_{36}]^{9-}$, EuW_{10}, Figure 1a) anion exhibits strong emission at room temperature with a decay time of millisecond order [22–25], and is often utilized to build up luminescent hybrid materials [18,19]. The EuW_{10} anion has two $W_5O_{18}^{6-}$ ligands that absorb ultraviolet (UV) light by O → W ligand-to-metal charge-transfer (LMCT). The intramolecular energy transfer from the LMCT state of $W_5O_{18}^{6-}$ ligands to the $^5D_0$ state of Eu^{3+} causes distinct orange-red light emission.

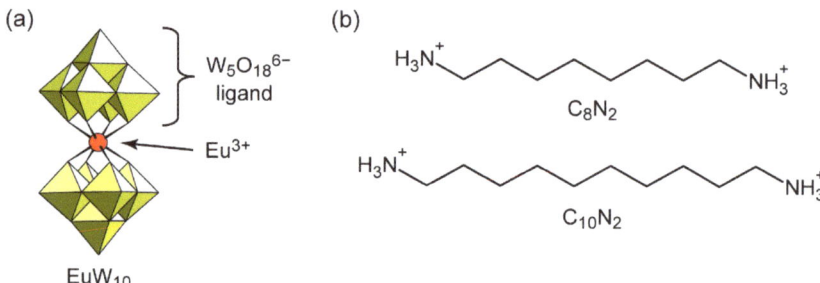

Figure 1. (a) Molecular structure of decatungstoeuropate anion, $[EuW_{10}O_{36}]^{9-}$ (EuW_{10}). Each polyhedron represents a WO_6 unit, and the red sphere represents a Eu^{3+} ion. (b) Molecular structure of bolaamphiphiles utilized in this work. Upper: 1,8-octamethylenediammonium, $[H_3N(CH_2)_8NH_3]^{2+}$ (C_8N_2); bottom: 1,10-decamethylenediammonium, $[H_3N(CH_2)_{10}NH_3]^{2+}$ ($C_{10}N_2$).

Cationic surfactants and polymer matrices as well as neutral block copolymers are effectively hybridized with the EuW_{10} anion to obtain luminescent nanocomposites [26–29], thin films [30–37], and sensors [38–41]. These systems are functional as soft matter and compatible with living organisms. However, single-crystalline ordering has not been achieved in most cases, which can be a drawback with the use as solid-state materials. The surfactant molecules are also effective as structure-directing reagents to construct one-dimensional tunnel and two-dimensional layer structures. Additionally, the EuW_{10} anion has rarely been crystallized with organic cations [42] and organic moieties [43,44].

Here, we report syntheses and structures of inorganic–organic hybrid crystals of the luminescent EuW_{10} anion and cationic bolaamphiphile surfactants, which have two hydrophilic heads in one molecule. The bolaamphiphile counterparts employed are 1,8-octamethylenediammonium ($[H_3N(CH_2)_8NH_3]^{2+}$, C_8N_2) and 1,10-decamethylenediammonium ($[H_3N(CH_2)_{10}NH_3]^{2+}$, $C_{10}N_2$), as shown in Figure 1b. Both hybrid crystals of C_8N_2-EuW_{10} and $C_{10}N_2$-EuW_{10} were successfully obtained as single crystals, and their crystal structures were unambiguously determined using X-ray diffraction measurements. The photoluminescence properties were evaluated by means of steady-state and time-resolved spectroscopy.

2. Results

2.1. Synthesis of EuW_{10}-Bolaamphiphile Hybrid Crystals

C_8N_2-EuW_{10} and $C_{10}N_2$-EuW_{10} hybrid crystals were synthesized via ion-exchange reactions using sodium salt of EuW_{10} (Na-EuW_{10}) and bolaamphiphile cations. The as-prepared precipitate of C_8N_2-EuW_{10} was obtained in 15–20% yield, and the as-prepared precipitate of $C_{10}N_2$-EuW_{10} was obtained in 40–50% yield. In each case, single crystals were successfully isolated from the synthetic filtrate after the removal of the as-prepared precipitate of C_8N_2-EuW_{10} or $C_{10}N_2$-EuW_{10}. The yields of isolated single crystals were ca. 50% for C_8N_2-EuW_{10}, and ca. 30% for $C_{10}N_2$-EuW_{10}. Figure 2 shows IR spectra of the as-prepared precipitates and single crystals of C_8N_2-EuW_{10} and $C_{10}N_2$-EuW_{10}. The spectra of C_8N_2-EuW_{10} (Figure 2b,c) showed characteristic peaks of the EuW_{10} anion in the range of 400–1000 cm^{-1} (935–945 cm^{-1} [ν_{as}(W=O$_t$)], 820–850 cm^{-1} [ν_{as}(W–O$_b$–W)], 700–710 cm^{-1} [ν_{as}(W–O$_c$–W)]) [45]. The peaks in the range of 2800–3000 cm^{-1} were derived from the C_8N_2 cation (2920 cm^{-1} [ν_{as}(−CH$_2$−)], 2850 cm^{-1} [ν_s(−CH$_2$−)]), which indicates the successful hybridization of the EuW_{10} anion and C_8N_2 cation. The IR spectra of $C_{10}N_2$-EuW_{10} (Figure 2d,e) also verified the formation of the $C_{10}N_2$-EuW_{10} hybrid crystal.

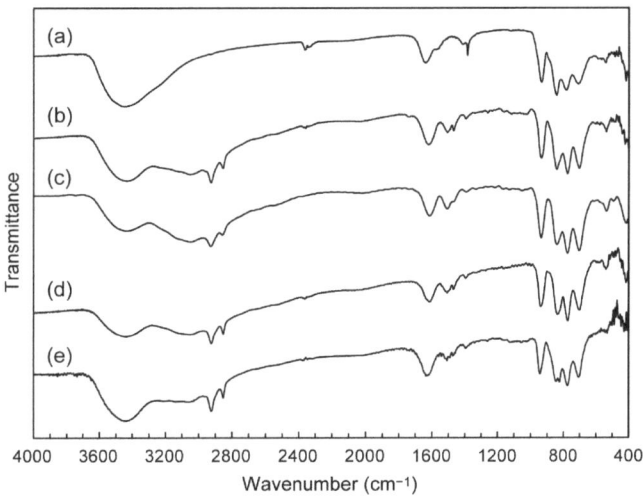

Figure 2. IR spectra of EuW_{10} and bolaamphiphile hybrid crystals: (**a**) starting material of Na-EuW_{10}; (**b**) as-prepared precipitate of C_8N_2-EuW_{10}; (**c**) single crystal of C_8N_2-EuW_{10}; (**d**) as-prepared precipitate of $C_{10}N_2$-EuW_{10}; and (**e**) single crystal of $C_{10}N_2$-EuW_{10}.

Figure 3 demonstrates powder XRD patterns of the C_8N_2-EuW_{10} and $C_{10}N_2$-EuW_{10} hybrid crystals. The XRD patterns of the C_8N_2-EuW_{10} as-prepared precipitate (Figure 3a) were crystalline, but slightly different from those of the C_8N_2-EuW_{10} single crystal (Figure 3b), and calculated from the results using single-crystal X-ray diffraction (Figure 3c). Slight differences in the peak position and intensity of the patterns may be derived from the desolvation of water molecules of crystallization (see below). The XRD pattern of the C_8N_2-EuW_{10} single crystal was similar to that calculated from results using single-crystal X-ray diffraction (Figure 3c). The XRD patterns of the $C_{10}N_2$-EuW_{10} as-prepared precipitate (Figure 3d) and single crystals (Figure 3e) were both similar to the calculated pattern from the results using single-crystal X-ray diffraction (Figure 3f). The results of IR spectra and powder XRD patterns indicate that both C_8N_2-EuW_{10} and $C_{10}N_2$-EuW_{10} hybrid crystals were obtained in a single phase and that the as-prepared precipitate and single crystal were essentially the same in their molecular and crystal structures.

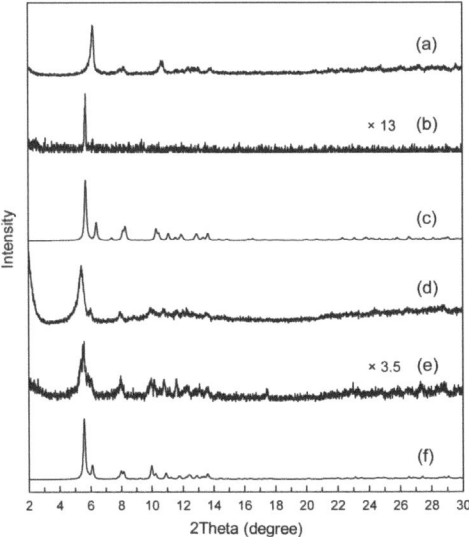

Figure 3. Powder X-ray diffraction patterns of EuW$_{10}$ and bolaamphiphile hybrid crystals: (**a**) as-prepared precipitate of C$_8$N$_2$-EuW$_{10}$; (**b**) single crystal of C$_8$N$_2$-EuW$_{10}$; (**c**) calculated pattern of C$_8$N$_2$-EuW$_{10}$ from the structure revealed using single-crystal X-ray diffraction; (**d**) as-prepared precipitate of C$_{10}$N$_2$-EuW$_{10}$; (**e**) single crystal of C$_{10}$N$_2$-EuW$_{10}$; and (**f**) calculated pattern of C$_{10}$N$_2$-EuW$_{10}$ from the structure revealed using single-crystal X-ray diffraction.

2.2. Crystal Structures of EuW$_{10}$-Bolaamphiphile Hybrid Crystals

The formulae of the hybrid crystals consisting of the EuW$_{10}$ anion and bolaamphiphile cations were revealed by means of single-crystal X-ray diffraction and CHN elemental analyses (Table 1). C$_8$N$_2$-EuW$_{10}$ has a formula of [H$_3$N(CH$_2$)$_8$NH$_3$]$_4$H[EuW$_{10}$O$_{36}$]·10H$_2$O, in which four C$_8$N$_2$ cations (2+ charge) and one H$^+$ (1+ charge) were connected to one EuW$_{10}$ anion (9−charge) due to charge compensation (Figure 4 and Figure S1). The presence of Na$^+$ was not detected using energy dispersive X-ray spectroscopy (EDS) analysis. Ten water molecules of crystallization were contained in the crystal lattice. As shown in the asymmetric unit (Figure S1), three crystallographically independent C$_8$N$_2$ cations (except for the C$_8$N$_2$ cation containing N1 and N2) were bent with gauche conformation. The associated H$^+$ was not observed using X-ray diffraction, but its presence was suggested by the bond valence sum (BVS) calculations [46]. The BVS value of plausibly protonated O atom (O21) in EuW$_{10}$ was 1.12, while those for other O atoms were 1.57–1.95. The Eu^{3+} cation held a distorted square-antiprismatic 8-fold coordination with Eu–O distances of 2.37–2.50 Å (mean value: 2.43 Å), and the shortest Eu···Eu distance was 10.49 Å, which is similar to those of Na-W$_{10}$ [25].

The crystal packing of C$_8$N$_2$-EuW$_{10}$ was a layer structure viewed along the *b*-axis (Figure 4a, left). The layer structure consisted of EuW$_{10}$ inorganic layers and C$_8$N$_2$ organic layers parallel to the *ab* plane with a periodicity of 15.5 Å. The crystal packing viewed along the *a*-axis exhibited a honeycomb-like feature (Figure 4a, right). The EuW$_{10}$ anions interacted with each other to form a one-dimensional chain structure (Figure 4b). The O···O distances were 2.73 Å (O12···O21) and 2.92 Å (O26···O28). As the BVS calculation suggested, the associated H$^+$ was located onto O21, and the short contact of O12···O21 (2.73 Å) was due to O–H···O hydrogen bonding [47]. Some water molecules (O37, O40, O43, and O44) were located inside the inorganic EuW$_{10}$ layer. These water molecules and EuW$_{10}$ anions formed a two-dimensional network (EuW$_{10}$-H$_2$O layer) through O–H···O hydrogen bonding with O···O distance ranging from 2.67 to 2.94 Å (mean value: 2.81 Å)

(Figure 4c). Some hydrophilic heads of C_8N_2 penetrated the EuW_{10}-H_2O layers with the N–H···O hydrogen bonding with distances of 2.71–3.04 Å (mean value: 2.85 Å) [47].

Table 1. Crystallographic data.

Compound	C_8N_2-EuW_{10}	$C_{10}N_2$-EuW_{10}
Chemical formula	$C_{32}H_{88}N_8EuW_{10}O_{46}$	$C_{35}H_{91}N_7EuW_{10}O_{42}$
Formula weight	3311.53	3272.59
Crystal system	monoclinic	monoclinic
Space group	$P2_1/c$ (No. 14)	$P2_1/c$ (No. 14)
a (Å)	15.5117 (6)	15.2827 (2)
b (Å)	15.3803 (5)	16.2555 (2)
c (Å)	31.3536 11)	32.0125 (6)
α (°)	90.0000	90.0000
β (°)	99.655 (4)	98.8002(15)
γ (°)	90.0000	90.0000
V (Å^3)	7374.2 (5)	7859.2 (2)
Z	4	4
ρ_{calcd} (g cm^{-3})	2.983	2.766
T (K)	93 (2)	93 (2)
Wavelength (Å)	0.71073	1.54184
μ (mm^{-1})	16.478	32.553
No. of reflections measured	100,838	65,068
No. of independent reflections	19,633	15,570
R_{int}	0.0911	0.0596
No. of parameters	852	535
R_1 ($I > 2\sigma(I)$)	0.0533	0.0716
wR_2 (all data)	0.1016	0.1980

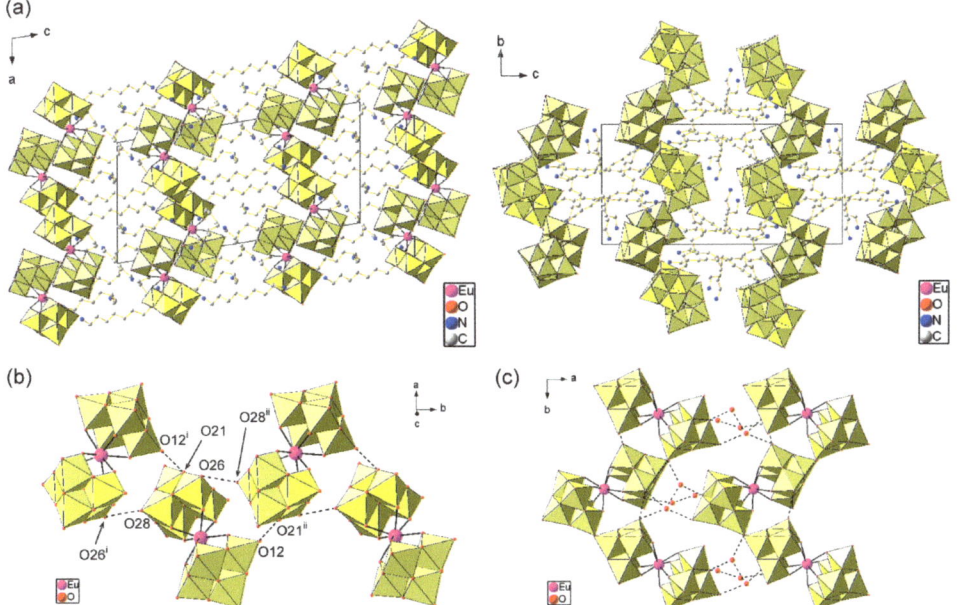

Figure 4. Crystal structure of C_8N_2-EuW_{10} (Eu: pink; C: gray; N: blue; O: red). WO_6 units in EuW_{10} are depicted in the polyhedral model. H atoms are omitted for clarity. (**a**) Packing diagram along *b*-axis (left) and *a*-axis (right). Solvent atoms are omitted for clarity. (**b**) One-dimensional arrangement

of EuW$_{10}$ anions. Broken lines represent short contacts between EuW$_{10}$ anions. Symmetry codes: (i) $2 - x$, $-0.5 + y$, $0.5 - z$; (ii) $2 - x$, $0.5 + y$, $0.5 - z$. (c) Molecular arrangement of the inorganic monolayer (*ab* plane). Broken lines represent short contacts between EuW$_{10}$ anions and solvents.

The chemical formula of C$_{10}$N$_2$-EuW$_{10}$ was determined to be [H$_3$N(CH$_2$)$_{10}$NH$_3$]$_{3.5}$H$_2$[EuW$_{10}$O$_{36}$]·6.5H$_2$O. Three and a half C$_{10}$N$_2$ cations (2+ charge) and two H$^+$ (1+ charge) were connected to one EuW$_{10}$ anion (9− charge) with four water molecules of crystallization (Figure 5). No residual Na$^+$ was observed using EDS analysis. As shown in the asymmetric unit (Figure S2), a half C$_{10}$N$_2$ cation (containing N7) was onto the inversion center with *anti*-conformation. Other C$_{10}$N$_2$ cations were bent with *gauche* conformation, and two C$_{10}$N$_2$ cations (with N3 and N4A, N4B; with N5 and N6A, N6B) were disordered with site occupancies of 0.558 and 0.442. Four water molecules were crystallographically assigned (Figure S2), while the presence of six and a half molecules per EuW$_{10}$ anion was suggested by the thermal gravimetric (TG) analyses (Figure S3). The associated H$^+$ was not detected using X-ray diffraction. The BVS value of O22 was 1.04 and seemed to be protonated (the BVS values of other O atoms: 1.59–1.96). The second H$^+$ was not revealed in its position but may be located in the vicinity of O18 (BVS value: 1.60) or O19 (BVS value: 1.59). The coordination environment was similar to those of C$_8$N$_2$-EuW$_{10}$ and Na-W$_{10}$ [25]: Eu–O distances of 2.43–2.47 Å (mean value: 2.45 Å) and the shortest Eu···Eu distance of 10.50 Å.

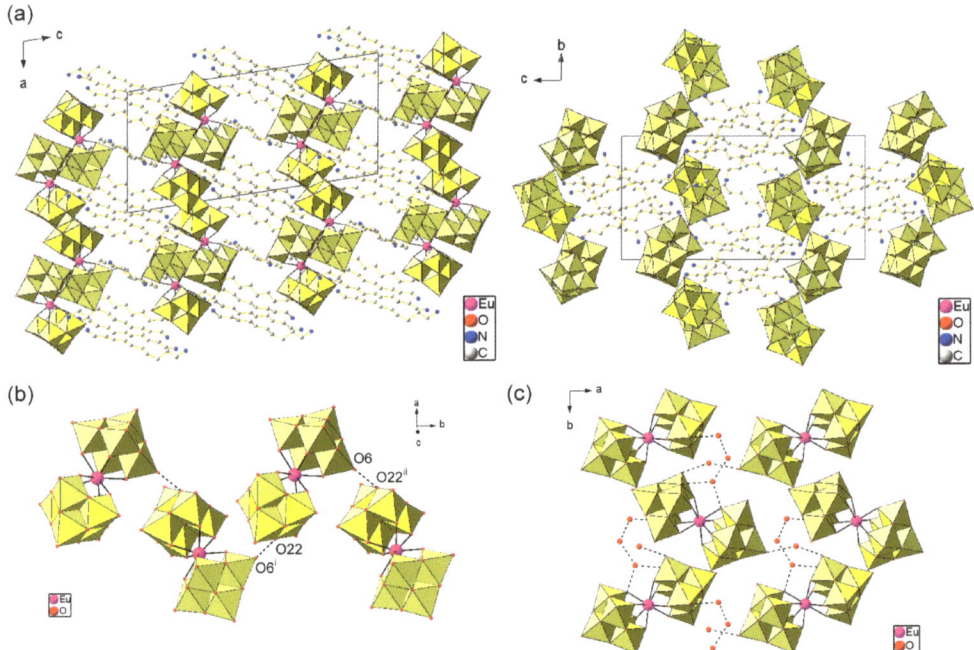

Figure 5. Crystal structure of C$_{10}$N$_2$-EuW$_{10}$ (Eu: pink; C: gray; N: blue; O: red). WO$_6$ units in EuW$_{10}$ are depicted in the polyhedral model. H atoms and disordered parts are omitted for clarity. (**a**) Packing diagram along the *b*-axis (left) and *a*-axis (right). Solvent atoms are omitted for clarity. (**b**) One-dimensional arrangement of EuW$_{10}$ anions. Broken lines represent short contacts between EuW$_{10}$ anions. Symmetry codes: (i) $1 - x$, $-0.5 + y$, $1.5 - z$; (ii) $1 - x$, $0.5 + y$, $1.5 - z$. (**c**) Molecular arrangement of the inorganic monolayer (*ab* plane). Broken lines represent short contacts between EuW$_{10}$ anions and solvents.

The crystal packing of C$_{10}$N$_2$-EuW$_{10}$ viewed along the *b*-axis was a layer structure composed of EuW$_{10}$ inorganic layers and C$_{10}$N$_2$ organic layers parallel to the *ab* plane (Figure 5a, left). The layered distance was 15.8 Å. As viewed along the *a*-axis (Figure 5a, right), the crystal packing was a honeycomb-like structure. The EuW_{10} anions formed a one-dimensional infinite chain (Figure 5b) by short contacts between O6 and O22 with an O···O distance of 2.71 Å. This short contact will be due to the O–H···O hydrogen bonding [47], since O22 is the plausibly protonated O atom by the BVS calculation. The crystallographically assigned water molecules were located inside the inorganic EuW_{10} layer to form a two-dimensional network with the EuW_{10} anions (EuW_{10}-H_2O layer) through the O–H···O hydrogen bonds (O···O distance: 2.72–3.04 Å; mean value: 2.88 Å) (Figure 5c). Some hydrophilic heads of $C_{10}N_2$ were located in the EuW_{10}-H_2O layers with N–H···O hydrogen bonds (N···O distance: 2.66–3.04 Å; mean value: 2.83 Å) [47].

2.3. Photoluminescent Properties of EuW_{10}-Bolaamphiphile Hybrid Crystals

The hybrid crystals of C_8N_2-EuW_{10} and $C_{10}N_2$-EuW_{10} exhibited distinct photoluminescence derived from the EuW_{10} anion. Figure 6 shows steady-state spectra of C_8N_2-EuW_{10} and $C_{10}N_2$-EuW_{10}. Diffuse reflectance spectra (Figure 6a) showed adsorptions around 395 nm and 465 nm, which were assigned as f-f transitions of Eu^{3+}: 395 nm for $^7F_0 \rightarrow {}^5L_6$ transition, and 465 nm for $^7F_0 \rightarrow {}^5D_2$ transition [14,15]. Each excitation spectrum (Figure 6b) exhibited a broad peak around 200–340 nm owing to the excitation into the O $\rightarrow$ W LMCT band in the $W_5O_{18}^{6-}$ ligands. The f-f transitions mentioned above were also observed in the excitation spectra. In the emission spectra, distinct peaks due to $^5D_0 \rightarrow {}^7F_J$ (J = 0, 1, 2, 3, 4) transition of Eu^{3+} were observed around 580–710 nm (Figure 5c) [22–25].

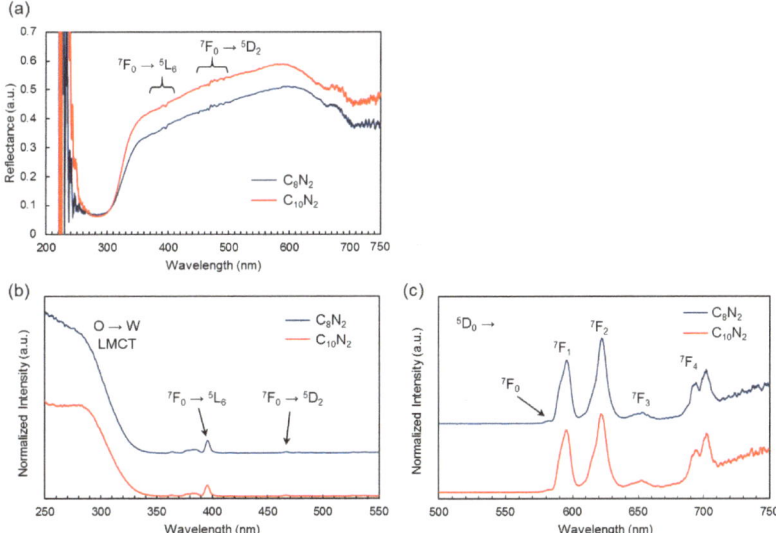

Figure 6. Steady-state spectra of C_8N_2-EuW_{10} and $C_{10}N_2$-EuW_{10}. The measurement temperature was 300 K: (**a**) diffuse reflectance spectra; (**b**) excitation spectra monitored on the emission at 595 nm; (**c**) emission spectra were measured with an excitation wavelength of 265 nm.

The photoluminescent properties of the C_8N_2-EuW_{10} and $C_{10}N_2$-EuW_{10} hybrid crystals were evaluated by means of time-resolved spectroscopy. The emission spectra acquired using a single pulse excitation (Figure 7a,b) exhibited characteristic emission derived from the EuW_{10} [22–25]. Emission peaks at 575 nm are assigned to $^5D_0 \rightarrow {}^7F_0$ transition, peaks at 587 and 593 nm to $^5D_0 \rightarrow {}^7F_1$, and peaks at 611 and 618 nm to $^5D_0 \rightarrow {}^7F_2$. The peaks around 650 nm are assignable to $^5D_0 \rightarrow {}^7F_3$ transition, and peaks at 691 and 700 nm to $^5D_0 \rightarrow {}^7F_4$ transition. The spectrum profiles of C_8N_2-EuW_{10} and $C_{10}N_2$-EuW_{10} were almost the same irrespective of the measurement temperatures. However, the emission decay profiles of C_8N_2-EuW$_{10}$ and $C_{10}N_2$-EuW$_{10}$ were different. The emission decay profiles of C_8N_2-EuW$_{10}$ were regarded with a single exponential function (blue plots in Figure 7c,d). The emission lifetimes were 3.0 ± 0.1 ms at 15 K and 2.5 ± 0.1 ms at 300 K (Table 2). In the case of $C_{10}N_2$-EuW$_{10}$, the emission decay profiles were approximated with two exponential functions (red plots in Figure 7c,d). The emission lifetimes at 15 K were estimated to be 0.94 ± 0.1 ms for a faster decay component and 3.1 ± 0.1 ms for a slower decay component (Table 2). The emission lifetimes at 300 K were 1.1 ± 0.1 ms and 1.8 ± 0.1 for a faster and slower component, respectively. The emission decay lifetimes of C_8N_2-EuW$_{10}$ and $C_{10}N_2$-EuW$_{10}$ at 15 K were comparable to that of Na-EuW$_{10}$ [25,33] but became shorter at 300 K. The increase in the number of carbon atoms in the bolaamphiphile cation resulted in a shorter emission lifetime at 300 K of $C_{10}N_2$-EuW$_{10}$ [33].

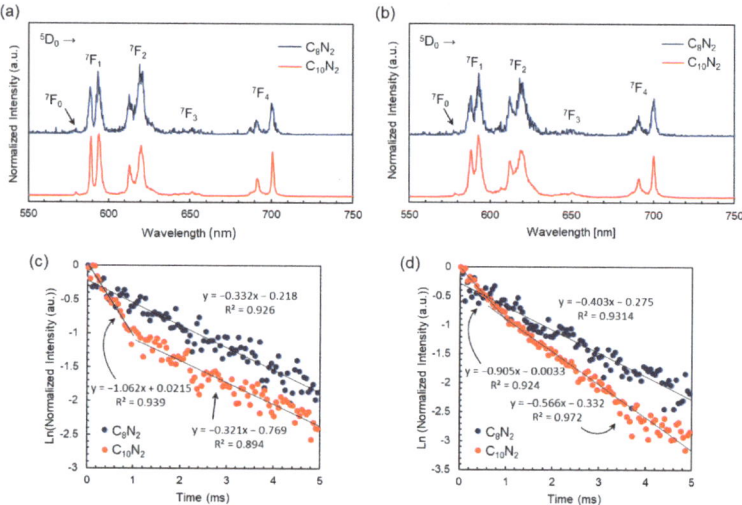

Figure 7. Photoluminescence properties of C_8N_2-EuW$_{10}$ and $C_{10}N_2$-EuW$_{10}$ investigated using time-resolved spectroscopy. Each spectrum or decay profile was obtained by a single pulse excitation with a wavelength of 266 nm. Emission spectra were acquired 50–100 μs after the excitation. Emission decays were monitored at the emission at 593 nm: (**a**) emission spectra measured at 15 K; (**b**) emission spectra measured at 300 K; (**c**) emission decay profiles measured at 15 K; (**d**) emission decay profiles measured at 300 K.

Table 2. Emission lifetimes (τ/ms) of C_8N_2-EuW$_{10}$ and $C_{10}N_2$-EuW$_{10}$.

Compound	15 K	300 K
C_8N_2-EuW$_{10}$	3.0 ± 0.1	2.5 ± 0.1
$C_{10}N_2$-EuW$_{10}$	$0.94 \pm 0.1 + 3.1 \pm 0.1$ [1]	$1.1 \pm 0.1 + 1.8 \pm 0.1$ [1]
Na-EuW$_{10}$	3.5 [2]	2.6 [3]

[1] Two exponential decays were applied. [2] The decay time at 4.2 K. Taken from Ref. [25] as a comparison. [3] The value at r.t. taken from Ref. [33].

As for the preparation of inorganic–organic luminescent materials, the lasing property is a promising character to be tackled in several applications. As shown in Figure 8, the emission intensity of C_8N_2-EuW$_{10}$ and $C_{10}N_2$-EuW$_{10}$ depended on the excitation laser power. After the threshold value of the excitation laser power, the emission intensity increased linearly, indicating the emergence of the lasing property [48]. The threshold values at 15 K were 28.5 and 25.4 mJ cm^{-2} for C_8N_2-EuW$_{10}$ and $C_{10}N_2$-EuW$_{10}$, respectively. The threshold values at 300 K were 26.0 and 25.2 mJ cm^{-2} for C_8N_2-EuW$_{10}$ and $C_{10}N_2$-EuW$_{10}$, respectively. These threshold values will be essentially in the same order.

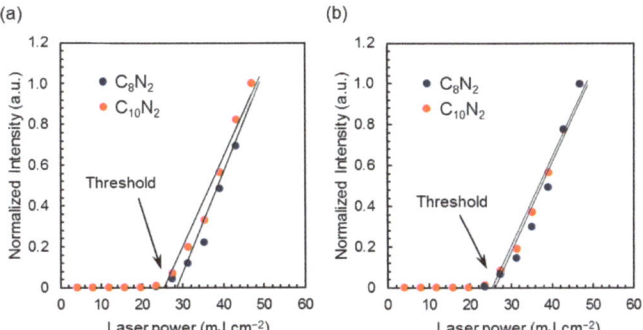

Figure 8. Emission intensity–excitation laser power dependency of C_8N_2-EuW_{10} and $C_{10}N_2$-EuW_{10} at (**a**) 15 K and (**b**) 300 K. Each data point was obtained by a single pulse excitation with a wavelength of 266 nm on the emission at 593 nm. Data acquisition time: 50–100 μs after the excitation.

3. Discussion

Lanthanide-containing polyoxometalate (Ln-POM) single crystals hybridized with surfactant molecules were first obtained in this work. Using bolaamphiphile surfactants was critical for the crystallization of the Ln-POM hybrid crystals. Bolaamphiphiles have two hydrophilic heads [49,50]. Hybrid crystals of POM with bolaamphiphiles have higher solubility in conventional solvents, and it is rather easier to isolate single crystals [51,52]. The size of the hydrophilic heads of C_8N_2 and $C_{10}N_2$ are smaller than those of quaternary alkylammonium cations, which may be another reason for the successful isolation of single crystals of C_8N_2-EuW_{10} and $C_{10}N_2$-EuW_{10}. The effect of surfactant length on luminescent properties will be an interesting topic; however, preparing single crystals with longer surfactants may be difficult.

The powder XRD patterns of the as-prepared precipitate (Figure 3a) and single crystal (Figure 3b) of the C_8N_2-EuW_{10} hybrid crystal were slightly different. On the other hand, the essential feature of the XRD patterns of as-prepared precipitate (Figure 3a) is similar to that calculated from the single-crystal structure of C_8N_2-EuW_{10} (Figure 3c). The as-prepared precipitate and single crystal of C_8N_2-EuW_{10} is considered to be the same phase. The differences in the peak position and intensity of the patterns will be derived from the desolvation of water molecules of crystallization, the different measurement temperatures (powder: room temperature; single crystal: 93 K), and the preferred orientation derived from the layered structure of C_8N_2-EuW_{10}. In the case of $C_{10}N_2$-EuW_{10}, the XRD patterns of the as-prepared precipitate (Figure 3d) and single crystal (Figure 3e) were quite similar. The water molecules in the $C_{10}N_2$-EuW_{10} hybrid crystal were located inside the inorganic layers of EuW_{10} with short-contact interaction, and plausibly less easily desorbed from the crystal lattice. TG analyses indicated the stability of C_8N_2-EuW_{10} and $C_{10}N_2$-EuW_{10} until 180–200 °C (Figure S3).

The structures of C_8N_2-EuW_{10} and $C_{10}N_2$-EuW_{10} hybrid crystals were unambiguously revealed by means of single-crystal X-ray diffraction measurements. In summary, the crystal structures were similar concerning the cell parameters (Table 1) and packing features (Figures 4 and 5). The crystal structures of C_8N_2-EuW_{10} and $C_{10}N_2$-EuW_{10} were layer structures viewed along the *b*-axis, and a honeycomb-like feature viewed along the *a*-axis. Such structural features are observed for some POM-surfactant crystals [52,53]. The packing features of EuW_{10} in C_8N_2-EuW_{10} and $C_{10}N_2$-EuW_{10} were almost the same, while the number of bolaamphiphile cations and their conformations were different. In both hybrid crystals, the EuW_{10} anions formed one-dimensional chain structures. The residual H^+ was relevant to the formation of the one-dimensional chain structure. These one-dimensional chains of EuW_{10} together with water molecules formed two-dimensional networks of EuW_{10}-H_2O parallel to the *ab* plane (Figures 4c and 5c).

The photoluminescence spectroscopy of C_8N_2-EuW_{10} and $C_{10}N_2$-EuW_{10} revealed the emission properties of C_8N_2-EuW_{10} and $C_{10}N_2$-EuW_{10}. The photoluminescence of the EuW_{10} anion was essentially retained: characteristic emission derived from Eu^{3+} (Figures 6 and 7) and an emission lifetime of millisecond order (Table 2). The hybridization of EuW_{10} with organic moieties sometimes shortens the emission lifetime (<1 ms) [33,54]; however, primary ammonium cation can retain the emission lifetime of millisecond order [31]. The primary ammonium cation can form N–H$\cdots$O hydrogen bonds between the EuW_{10} anion to prevent water molecules from approaching near Eu^{3+}. The excitation energy owing to O → W LMCT can transfer to Eu^{3+} without nonradiative deactivation through the vibration states of the high-frequency O–H oscillators of water molecules [55,56]. The EuW_{10} anion has no coordinated water and therefore a long emission decay time (3.5 ms at 4.2 K) and high quantum yield (0.99) for Na-EuW_{10} at 4.2 K [25]. As shown in Table 2, the emission decay time of C_8N_2-EuW_{10} (3.0 ± 0.1 ms) and $C_{10}N_2$-EuW_{10} (3.1 ± 0.1 ms for a slower component) at 15 K were comparable to that of Na-EuW_{10} (3.5 ms) at 4.2 K. This implies that the emission behavior of C_8N_2-EuW_{10} and $C_{10}N_2$-EuW_{10} was almost identical to that of Na-EuW_{10} derived from the suppression of the thermal deactivation of the excitation energy at the low temperature. At the high temperature (300 K), the emission decay time of C_8N_2-EuW_{10} (2.5 ± 0.1 ms) was similar to that of Na-EuW_{10} (2.6 ms), but $C_{10}N_2$-EuW_{10} exhibited the faster decay time of 1.8 ± 0.1 ms. This will be due to more carbon atoms in the crystal lattice [33] and fewer N–H$\cdots$O hydrogen bonds between EuW_{10} and surfactant cations. The kinetic constants of energy transfer can be estimated using the magnetic-dipole $^5D_0 \to {}^7F_1$ transition as a standard [56], since the rate of $^5D_0 \to {}^7F_1$ (1.35×10^2 s^{-1}) is almost independent of the geometry of Eu^{3+}. The relative intensity of $^5D_0 \to {}^7F_1$ emission to the total emission ($^5D_0 \to {}^7F_n$, n = 0–4) at 300 K was 0.38 for C_8N_2-EuW_{10} and 0.41 for $C_{10}N_2$-EuW_{10}, respectively. Therefore, for C_8N_2-EuW_{10}, the radiative rate (k_{rad}) was 3.6×10^2 (= $1.35 \times 10^2/0.36$) s^{-1} and the experimental decay rate was $(4.0 \pm 0.2) \times 10^2$ (= $1/((2.5 \pm 0.1) \times 10^{-3})$) s^{-1}, and then the estimated nonradiative rate (k_{nr}) was $(0.4 \pm 0.2) \times 10^2$ (= $(4.0 \pm 0.2) \times 10^2 - 3.6 \times 10^2$) s^{-1}. For $C_{10}N_2$-EuW_{10}, the respective values of radiative rate (k_{rad}) and experimental decay rate were 3.3×10^2 (= $1.35 \times 10^2/0.41$) s^{-1} and $(5.6 \pm 0.3) \times 10^2$ (= $1/((1.9 \pm 0.1) \times 10^{-3})$) s^{-1}, and the estimated nonradiative rate (k_{nr}) was $(2.3 \pm 0.3) \times 10^2$ (= $(5.6 \pm 0.3) \times 10^2 - 3.3 \times 10^2$) s^{-1}. The reason for the presence of fast decay components in the $C_{10}N_2$-EuW_{10} emission (Table 2) was unclear but may be derived from the presence of more H^+ in the crystal lattice. In addition, both C_8N_2-EuW_{10} and $C_{10}N_2$-EuW_{10} hybrid crystals exhibited lasing properties (Figure 8). The threshold values (25–28 mJ cm^{-2}) were larger than those of recent organic lasers [57,58]. Although further improvement in materials processing will be necessary, the photoluminescence properties of C_8N_2-EuW_{10} and $C_{10}N_2$-EuW_{10} mentioned above show the possibility of a new series of inorganic–organic hybrid phosphors.

4. Materials and Methods

4.1. Materials

Chemical reagents purchased from commercial sources (FUJIFILM Wako Pure Chemical Corporation, Osaka, Japan, Tokyo Chemical Industry Co., Ltd. (TCI), Tokyo, Japan and Kanto Chemical Co., Inc., Tokyo, Japan) were utilized without further purification. Solid 1,8-octamethylenediammonium chloride ([$H_3N(CH_2)_8NH_3$]Cl_2, C_8N_2-Cl) and 1,10-decamethylenediammonium chloride ([$H_3N(CH_2)_{10}NH_3$]Cl_2, $C_{10}N_2$-Cl) were prepared by adding equimolar hydrochloric acid to 1,8-octanediamine and 1,10-decanediamine, respectively. The sodium salt of EuW_{10} ($Na_9[EuW_{10}O_{36}]\cdot 32H_2O$, Na-$EuW_{10}$) was prepared according to the literature [25].

4.2. Measurements

Infrared (IR) spectra were recorded with an FT/IR-4200ST spectrometer (Jasco Corporation, Tokyo, Japan, KBr pellet method). Powder X-ray diffraction (XRD) patterns were measured on a MiniFlex300 diffractometer (Rigaku Corporation, Tokyo, Japan, Cu Kα radiation, λ = 1.54056 Å). CHN (carbon, hydrogen, and nitrogen) elemental analyses were performed with a 2400II elemental analyzer (PerkinElmer, Inc., Waltham, MA, USA). Energy dispersive X-ray (EDS) spectroscopy was performed on a JSM-6000Plus (JEOL, Tokyo, Japan). Thermal gravimetric (TG) analyses were measured with a TG/DTA-6200 (Seiko Instruments, Chiba, Japan) at a heating rate of 10 $°C$ min^{-1} under a nitrogen atmosphere.

Steady-state spectra (diffuse-reflectance, excitation, and emission) were obtained at 300 K on an FP-6500 fluorescence spectrometer (Jasco Corporation, Tokyo, Japan) using Xe lamp excitation. Time-resolved emission spectra were acquired at 15 and 300 K, using an Ultra CFR 400 YAG:Nd^{3+} laser (Big Sky Laser Technologies, Inc., Bozeman, MT, USA, 266 nm fourth harmonics, pulse duration 10 ns with a repetition rate of 10 Hz) as an excitation source. A Spectra Pro 2300i and PI-Max intensified CCD camera (Princeton Instruments, Inc., Trenton, NJ, USA) were employed as a spectrometer and a detector, respectively. Pelletized samples of the as-prepared precipitate of C_8N_2-EuW$_{10}$ and $C_{10}N_2$-EuW$_{10}$ were utilized for the photoluminescence measurements.

4.3. Synthesis of C_8N_2-EuW$_{10}$ Hybrid Crystal

A water/ethanol (20 mL, 1:1 (v/v)) solution of C_8N_2-Cl (0.11 g, 0.50 mmol) was added to an aqueous solution (20 mL) of Na-EuW$_{10}$ (0.47 g, 0.14 mmol), and stirred for 10 min. The resultant suspension was heated until 60 $°C$ with stirring (for 5–10 min) and quickly filtrated to obtain a colorless as-prepared precipitate of C_8N_2-EuW$_{10}$ (0.078 g, yield 17%). Colorless plates of C_8N_2-EuW$_{10}$ single crystal were isolated from the hot synthetic filtrate kept at 25–42 $°C$ (0.25 g, yield 51%). No presence of Na$^+$ in the C_8N_2-EuW$_{10}$ single crystals was confirmed using EDS spectroscopy. Anal. Calcd for $C_{32}H_{105}N_8EuW_{10}O_{44}$: C, 11.66; H, 3.21; N, 3.40%. Found: C, 11.45; H, 2.98; N, 3.29%. IR (KBr disk): 936 (m), 840 (s), 775 (s), 704 (s), 537 (w), 489 (w), 440 (w), 423 (m) cm^{-1}.

4.4. Synthesis of $C_{10}N_2$-EuW$_{10}$ Hybrid Crystal

The synthesis of $C_{10}N_2$-EuW$_{10}$ was carried out using a similar procedure as for C_8N_2-EuW$_{10}$. A colorless as-prepared precipitate of $C_{10}N_2$-EuW$_{10}$ was obtained from the combined suspension of Na-EuW$_{10}$ and $C_{10}N_2$-Cl (0.22 g, yield 49%). Colorless plate single crystals of $C_{10}N_2$-EuW$_{10}$ were isolated from the hot synthetic filtrate (0.16 g, yield 34%). No presence of Na$^+$ in the $C_{10}N_2$-EuW$_{10}$ single crystals was confirmed using EDS spectroscopy. Anal. Calcd for $C_{37}H_{99}N_7EuW_{10}O_{37}$: C, 13.65; H, 3.06; N, 3.01%. Found: C, 13.79; H, 3.15; N, 3.14%. IR (KBr disk): 946 (m), 848 (s), 824 (m), 773 (s), 703 (s), 531 (w), 494 (w), 459 (w), 418 (m) cm^{-1}.

4.5. X-ray Crystallography

Single-crystal X-ray diffraction measurements were performed with a Rigaku XtaLAB PRO P200 diffractometer using graphite monochromated Mo Kα radiation (λ = 0.71073 Å) or Cu Kα radiation (λ = 1.54184 Å). The data collection and processing including absorption correction were performed by CrysAlisPro (Version 1.171.39.46) [59]. Crystal structures were solved by SHELXT (Version 2018/2) [60], and refined through the full-matrix least-squares using SHELXL (Version 2018/3) [61]. The diffraction data recorded at the 2D beamline in the Pohang Accelerator Laboratory (PAL) confirmed the same crystal structure. CCDC 2352151-2352152.

5. Conclusions

Lanthanide-containing polyoxometalate-surfactant hybrid crystals were first obtained as single crystals. A highly luminescent decatungstoeuropate (EuW$_{10}$) anion was successfully crystallized with bolaamphiphile surfactant cations (C_8N_2 and $C_{10}N_2$). Both C_8N_2-EuW$_{10}$ and $C_{10}N_2$-EuW$_{10}$ hybrid crystals had a similar packing of the EuW$_{10}$ anion: a layer structure viewed along the *b*-axis and a honeycomb-like structure viewed along the *a*-axis. The EuW$_{10}$ anions formed a two-dimensional network parallel to the *ab* plane by O–H···O hydrogen bonding with water molecules. The luminescent properties of C_8N_2-

EuW_{10} and $C_{10}N_2$-EuW_{10} were investigated by means of steady-state and time-resolved spectroscopy. The characteristic emission owing to EuW_{10} was essentially retained after the hybrid crystals. The emission decay time of $C_{10}N_2$-EuW_{10} became shorter than that of $C_{10}N_2$-EuW_{10}, especially at a high temperature (300 K), suggesting the thermal deactivation of the excitation energy derived from the longer organic surfactant of $C_{10}N_2$. The C_8N_2-EuW_{10} and $C_{10}N_2$-EuW_{10} hybrid crystals exhibited preliminary lasing properties, which is promising as a new category of inorganic–organic phosphors.

Supplementary Materials: The following supporting information can be downloaded at: https://www.mdpi.com/article/10.3390/inorganics12060146/s1, Figure S1: Asymmetric unit of C_8N_2-EuW_{10}; Figure S2: Asymmetric unit of $C_{10}N_2$-EuW_{10}; Figure S3: TG profiles of C_8N_2-EuW_{10} and $C_{10}N_2$-EuW_{10}.

Author Contributions: Conceptualization, T.I.; methodology, R.I. and T.S.; validation, Y.S., T.K. and T.I.; formal analysis, T.I., Y.S., T.K., T.M. and Y.O.; investigation, R.I., R.K., Y.S., T.S. and K.K.; resources, Y.S. and Y.O.; writing—original draft preparation, T.I.; writing—review and editing, T.I. and Y.S.; visualization, T.I., Y.S., R.I. and T.M.; funding acquisition, T.I. All authors have read and agreed to the published version of the manuscript.

Funding: This research was funded in part by JSPS KAKENHI (grant number JP21K05232), and the Research and Study Project of Tokai University Research Organization.

Data Availability Statement: Further details of the crystal structure investigation (CCDC 2352151-2352152) can be obtained free of charge via www.ccdc.cam.ac.uk/data_request/cif (accessed on 30 April 2024), or by emailing data_request@ccdc.cam.ac.uk, or by contacting The Cambridge Crystallographic Data Centre, 12 Union Road, Cambridge CB2 1EZ, UK; Fax: +44-1223-336033.

Acknowledgments: This work is partially supported by the Tokai University Imaging Center for Advanced Research. X-ray diffraction measurements with synchrotron radiation were performed at the Pohang Accelerator Laboratory (Beamline 2D, proposal No. 2019-1st-2D-015), a synchrotron radiation facility in Pohang, Republic of Korea, supported by Pohang University of Science and Technology (POSTECH).

Conflicts of Interest: The authors declare no conflicts of interest. The funders had no role in the design of the study; in the collection, analyses, or interpretation of data; in the writing of the manuscript; or in the decision to publish the results.

References

1. Adachi, G.; Imanaka, N.; Tamura, S. Ionic Conducting Lanthanide Oxides. *Chem. Rev.* **2002**, *102*, 2405–2429. [CrossRef] [PubMed]
2. Coronado, E. Molecular magnetism: From chemical design to spin control in molecules, materials and devices. *Nat. Rev. Mater.* **2020**, *5*, 87–104. [CrossRef]
3. Aiba, Y.; Sumaoka, J.; Komiyama, M. Artificial DNA cutters for DNA manipulation and genome engineering. *Chem. Soc. Rev.* **2011**, *40*, 5657–5668. [CrossRef] [PubMed]
4. Eliseeva, S.V.; Bünzli, J.-C.G. Lanthanide luminescence for functional materials and bio-sciences. *Chem. Soc. Rev.* **2010**, *39*, 189–227. [CrossRef] [PubMed]
5. Carlos, L.D.; Ferreira, R.A.S.; de Zea Bermudez, V.; Julian-Lopez, B.; Escribano, P. Progress on lanthanide-based organic–inorganic hybrid phosphors. *Chem. Soc. Rev.* **2011**, *40*, 536–549. [CrossRef] [PubMed]
6. Heine, J.; Müller-Buschbaum, K. Engineering metal-based luminescence in coordination polymers and metal–organic frameworks. *Chem. Soc. Rev.* **2013**, *42*, 9232–9242. [CrossRef] [PubMed]
7. Hasegawa, Y.; Nakanishi, T. Luminescent lanthanide coordination polymers for photonic applications. *RSC Adv.* **2015**, *5*, 338–353. [CrossRef]
8. Binnemans, K. Lanthanide-based luminescent hybrid materials. *Chem. Rev.* **2009**, *109*, 4283–4374.
9. Zhang, R.; Shang, J.; Xin, J.; Xie, B.; Li, Y.; Möhwald, H. Self-assemblies of luminescent rare earth compounds in capsules and multilayers. *Adv. Colloid Interface Sci.* **2014**, *207*, 361–375. [CrossRef]
10. Yamase, T. Photo- and electrochromism of polyoxometalates and related materials. *Chem. Rev.* **1998**, *98*, 307–325. [CrossRef]
11. Pope, M.T. Polyoxometalates. In *Handbook on the Physics and Chemistry of Rare Earth*; Gschneidner, K.A., Jr., Bunzli, J.-C.G., Pecharsky, V.K., Eds.; Elsevier: Amsterdam, The Netherlands, 2008; Volume 38, pp. 337–382.
12. Zhao, J.-W.; Li, Y.-Z.; Chen, L.-J.; Yang, G.-Y. Research progress on polyoxometalate-based transition-metal–rare-earth heterometallic derived materials: Synthetic strategies, structural overview and functional applications. *Chem. Commun.* **2016**, *52*, 4418–4445. [CrossRef]

13. Boskovic, C. Rare Earth Polyoxometalates. *Acc. Chem. Res.* **2017**, *50*, 2205–2214. [CrossRef] [PubMed]
14. Yamase, T.; Naruke, H.; Sasaki, Y. Crystallographic characterization of the polyoxotungstate [Eu$_3$(H$_2$O)$_3$(SbW$_9$O$_{33}$)(W$_5$O$_{18}$)$_3$]$^{18-}$ and energy transfer in its crystalline lattices. *J. Chem. Soc. Dalton Trans.* **1990**, *5*, 1687–1696. [CrossRef]
15. Yamase, T.; Naruke, H. X-ray structural and photoluminescence spectroscopic investigation of the europium octamolybdate polymer Eu$_2$(H$_2$O)$_{12}$[Mo$_8$O$_{27}$]·6H$_2$O and intramolecular energy transfer in the crystalline lattice. *J. Chem. Soc. Dalton Trans.* **1991**, *2*, 285–292. [CrossRef]
16. Tewari, S.; Adnan, M.; Balendra; Kumar, V.; Jangra, G.; Prakash, G.V.; Ramanan, A. Photoluminescence properties of two closely related isostructural series based on Anderson-Evans cluster coordinated with lanthanides [Ln(H$_2$O)$_7${X(OH)$_6$Mo$_6$O$_{18}$}]·yH$_2$O, X = Al, Cr. *Front. Chem.* **2019**, *6*, 631. [CrossRef] [PubMed]
17. Shitamatsu, K.; Kojima, T.; Waddell, P.G.; Sugiarto; Ooyama, H.E.; Errington, R.J.; Sadakane, M. Structural characterization of cerium-encapsulated Preyssler-type phosphotungstate: Additional evidence of Ce(III) in the cavity. *Z. Anorg. Allg. Chem.* **2021**, *647*, 1239–1244. [CrossRef]
18. Qi, W.; Wu, L. Polyoxometalate/polymer hybrid materials: Fabrication and properties. *Polym. Int.* **2009**, *58*, 1217–1225. [CrossRef]
19. Granadeiro, C.M.; de Castro, B.; Balula, S.S.; Cunha-Silva, L. Lanthanopolyoxometalates: From the structure of polyanions to the design of functional materials. *Polyhedron* **2013**, *52*, 10–24. [CrossRef]
20. Ritchie, C.; Baslon, V.; Moore, E.G.; Reber, C.; Boskovic, C. Sensitization of lanthanoid luminescence by organic and inorganic ligands in lanthanoid-organic-polyoxometalates. *Inorg. Chem.* **2012**, *51*, 1142–1151. [CrossRef]
21. Wu, H.; Zhi, M.; Chen, H.; Singh, V.; Ma, P.; Wang, J.; Niu, J. Well-tuned white-light-emitting behaviours in multicenter-Ln polyoxometalate derivatives: A photoluminescence property and energy transfer pathway study. *Spectrochim. Acta Part A* **2019**, *223*, 117294. [CrossRef]
22. Stillman, M.J.; Thomson, A.J. Emission spectra of some lanthanoid decatungstate and undecatungstosilicate ions. *J. Chem. Soc. Dalton Trans.* **1976**, *12*, 1138–1144. [CrossRef]
23. Blasse, G.; Dirksen, G.J.; Zonnevijlle, F. The luminescence of some lanthanide decatungstates and other polytungstates. *J. Inorg. Nucl. Chem.* **1981**, *43*, 2847–2853. [CrossRef]
24. Ballardini, R.; Mulazzani, Q.G.; Venturi, M.; Bolletta, F.; Balzani, V. Photophysical characterization of the decatungstoeuropate(9−) anion. *Inorg. Chem.* **1984**, *23*, 300–305. [CrossRef]
25. Sugeta, M.; Yamase, T. Crystal structure and luminescence site of Na$_9$[EuW$_{10}$O$_{36}$]·32H$_2$O. *Bull. Chem. Soc. Jpn.* **1993**, *66*, 444–449. [CrossRef]
26. Li, W.; Qi, W.; Li, W.; Sun, H.; Bu, W.; Wu, L. A highly transparent and luminescent hybrid based on the copolymerization of surfactant-encapsulated polyoxometalate and methyl methacrylate. *Adv. Mater.* **2005**, *17*, 2688–2692. [CrossRef]
27. Wang, Z.; Zhang, R.; Ma, Y.; Peng, A.; Fu, H.; Yao, J. Chemically responsive luminescent switching in transparent flexible self-supporting [EuW$_{10}$O$_{36}$]$^{9-}$-agarose nanocomposite thin films. *J. Mater. Chem.* **2010**, *20*, 271–277. [CrossRef]
28. Zhang, J.; Liu, Y.; Li, Y.; Zhao, H.; Wan, X. Hybrid assemblies of Eu-containing polyoxometalates and hydrophilic block copolymers with enhanced emission in aqueous solution. *Angew. Chem. Int. Ed.* **2012**, *51*, 4598–4602. [CrossRef]
29. Pinto, R.J.B.; Granadeiro, C.M.; Freire, C.S.R.; Silvestre, A.J.D.; Pascoal Neto, C.; Ferreira, R.A.S.; Carlos, L.D.; Cavaleiro, A.M.V.; Trindade, T.; Nogueira, H.I.S. Luminescent transparent composite films based on lanthanopolyoxometalates and filmogenic polysaccharides. *Eur. J. Inorg. Chem.* **2013**, *2013*, 1890–1896. [CrossRef]
30. Wang, J.; Wang, H.; Fu, L.; Liu, F.; Zhang, H. Study on highly ordered luminescent Langmuir–Blodgett films of heteropolytungstate complexes containing lanthanide. *Thin Solid Films* **2002**, *415*, 242–247. [CrossRef]
31. Wang, J.; Wang, H.; Wang, Z.; Yin, Y.; Liu, F.; Li, H.; Fu, L.; Zhang, H. Luminescent hybrid Langmuir–Blodgett films of polyoxometaloeuropate. *J. Alloys Compd.* **2004**, *365*, 102–107. [CrossRef]
32. Jiang, M.; Zhai, X.; Liu, M. Fabrication and photoluminescence of hybrid organized molecular films of a series of gemini amphiphiles and europium(III)-containing polyoxometalate. *Langmuir* **2005**, *21*, 11128–11135. [CrossRef] [PubMed]
33. Ito, T.; Yashiro, H.; Yamase, T. Two-dimensional photoluminescence behavior of Langmuir-Blodgett monolayers and multilayers composed of decatungstoeuropate. *J. Cluster Sci.* **2006**, *17*, 375–387. [CrossRef]
34. Clemente-León, M.; Coronado, E.; López-Muñoz, A.; Repetto, D.; Ito, T.; Konya, T.; Yamase, T.; Constable, E.C.; Housecroft, C.E.; Doyle, K.; et al. Dual-emissive photoluminescent Langmuir-Blodgett films of decatungstoeuropate and an amphiphilic iridium complex. *Langmuir* **2010**, *26*, 1316–1324. [CrossRef] [PubMed]
35. Zhang, T.R.; Lu, R.; Zhang, H.Y.; Xue, P.C.; Feng, W.; Liu, X.L.; Zhao, B.; Zhao, Y.Y.; Li, T.J.; Yao, J.N. Highly ordered photoluminescent self-assembled films based on polyoxotungstoeuropate complex Na$_9$[EuW$_{10}$O$_{36}$]. *J. Mater. Chem.* **2003**, *13*, 580–584. [CrossRef]
36. Wang, Y.; Wang, X.; Hu, C.; Shi, C. Photoluminescent organic–inorganic composite films layer-by-layer self-assembled from the rare-earth-containing polyoxometalate Na$_9$[EuW$_{10}$O$_{36}$] and poly(allylamine hydrochloride). *J. Mater. Chem.* **2002**, *12*, 703–707. [CrossRef]
37. Tang, P.; Hao, J. Photoluminescent honeycomb films templated by microwater droplets. *Langmuir* **2010**, *26*, 3843–3847. [CrossRef] [PubMed]
38. Gao, J.; Zhang, Y.; Jia, G.; Jiang, Z.; Wang, S.; Lu, H.; Song, B.; Li, C. A direct imaging of amphiphilic catalysts assembled at the interface of emulsion droplets using fluorescence microscopy. *Chem. Commun.* **2008**, *3*, 332–334. [CrossRef] [PubMed]

39. Guo, Y.; Gong, Y.; Gao, Y.; Xiao, J.; Wang, T.; Yu, L. Multi-stimuli responsive supramolecular structures based on azobenzene surfactant-encapsulated polyoxometalate. *Langmuir* **2016**, *32*, 9293–9300. [CrossRef]
40. Xia, C.; Zhang, S.; Tan, Y.; Sun, D.; Sun, P.; Cheng, X.; Xin, X. Self-assembly of europium-containing polyoxometalates/tetra-*n*-alkyl ammonium with enhanced emission for Cu^{2+} detection. *ACS Omega* **2018**, *3*, 14953–14961. [CrossRef]
41. Liu, H.; Lv, Y.; Li, S.; Yang, F.; Liu, S.; Wang, C.; Sun, J.-Q.; Meng, H.; Gao, G.-G. A solar ultraviolet sensor based on fluorescent polyoxometalate and viologen. *J. Mater. Chem. C* **2017**, *5*, 9383–9388. [CrossRef]
42. Shen, D.-F.; Li, S.; Liu, H.; Jiang, W.; Zhang, Q.; Gao, G.-G. A durable and fast-responsive photochromic and switchable luminescent polyviologen–polyoxometalate hybrid. *J. Mater. Chem. C* **2015**, *3*, 12090–12097. [CrossRef]
43. Song, C.-Y.; Chai, D.-F.; Zhang, R.-R.; Liu, H.; Qiu, Y.-F.; Guo, H.-D.; Gao, G.-G. A silver-alkynyl cluster encapsulating a fluorescent polyoxometalate core: Enhanced emission and fluorescence modulation. *Dalton Trans.* **2015**, *44*, 3997–4002. [CrossRef] [PubMed]
44. Zhang, S.-S.; Su, H.-F.; Wang, Z.; Wang, X.-P.; Chen, W.-X.; Zhao, Q.-Q.; Tung, C.-H.; Sun, D.; Zheng, L.-S. Elimination-fusion self-assembly of a nanometer-scale 72-nucleus silver cluster caging a pair of $[EuW_{10}O_{36}]^{9-}$ polyoxometalates. *Chem. Eur. J.* **2018**, *24*, 1998–2003. [CrossRef]
45. Lis, S. Applications of spectroscopic methods in studies of polyoxometalates and their complexes with lanthanide(III) ions. *J. Alloys Compd.* **2000**, *300–301*, 88–94. [CrossRef]
46. Brown, I.D.; Altermatt, D. Bond-valence parameters obtained from a systematic analysis of the inorganic crystal structure database. *Acta Crystallogr. Sect. B* **1985**, *41*, 244–247. [CrossRef]
47. Desiraju, G.R.; Steiner, T. *The Weak Hydrogen Bond in Structural Chemistry and Biology*; Oxford University Press: New York, NY, USA, 1999; pp. 12–16.
48. Svelto, O. *Principles of Lasers*, 5th ed.; Springer: New York, NY, USA, 2010; Chapter 7.
49. Fuhrhop, J.-H.; Wang, T. Bolaamphiphiles. *Chem. Rev.* **2004**, *104*, 2901–2937. [CrossRef]
50. Shimizu, T.; Masuda, M.; Minamikawa, H. Supramolecular nanotube architectures based on amphiphilic molecules. *Chem. Rev.* **2005**, *105*, 1401–1443. [CrossRef] [PubMed]
51. Kiyota, Y.; Kojima, T.; Kawahara, R.; Taira, M.; Naruke, H.; Kawano, M.; Uchida, S.; Ito, T. Porous layered inorganic-organic hybrid frameworks constructed from polyoxovanadate and bolaamphiphiles. *Cryst. Growth Des.* **2021**, *21*, 7230–7239. [CrossRef]
52. Ikuma, H.; Aoki, S.; Kawahara, K.; Ono, S.; Iwamatsu, H.; Kobayashi, J.; Kiyota, Y.; Okamura, Y.; Higuchi, M.; Ito, T. An inorganic–organic hybrid framework composed of polyoxotungstate and long-chained bolaamphiphile. *Int. J. Mol. Sci.* **2023**, *24*, 2824. [CrossRef]
53. Misawa, T.; Kobayashi, J.; Kiyota, Y.; Watanabe, M.; Ono, S.; Okamura, Y.; Koguchi, S.; Higuchi, M.; Nagase, Y.; Ito, T. Dimensional control in polyoxometalate crystals hybridized with amphiphilic polymerizable ionic liquids. *Materials* **2019**, *12*, 2283. [CrossRef]
54. Mihara, A.; Kojima, T.; Suda, Y.; Maezawa, K.; Sumi, T.; Mizoe, N.; Watanabe, A.; Iwamatsu, H.; Oda, Y.; Okamura, Y.; et al. Photoluminescent layered crystal consisting of Anderson-type polyoxometalate and surfactant toward a potential inorganic–organic hybrid laser. *Int. J. Mol. Sci.* **2024**, *25*, 345. [CrossRef]
55. Horrocks, W.D., Jr.; Sudnick, D.R. Lanthanide ion luminescence probes of the structure of biological macromolecules. *Acc. Chem. Res.* **1981**, *14*, 384–392. [CrossRef]
56. Yamase, T.; Kobayashi, T.; Sugeta, M.; Naruke, H. Europium(III) luminescence and intramolecular energy transfer studies of polyoxometalloeuropates. *J. Phys. Chem. A* **1997**, *101*, 5046–5053. [CrossRef]
57. Samuel, I.D.W.; Turnbull, G.A. Organic semiconductor lasers. *Chem. Rev.* **2007**, *107*, 1272–1295. [CrossRef] [PubMed]
58. Kuehne, A.J.C.; Gather, M.C. Organic lasers: Recent developments on materials, device geometries, and fabrication techniques. *Chem. Rev.* **2016**, *116*, 12823–12864. [CrossRef] [PubMed]
59. *CrysAlisPro*; Rigaku Oxford Diffraction: Tokyo, Japan, 2015.
60. Sheldrick, G.M. SHELXT—Integrated space-group and crystal structure determination. *Acta Crystallogr. Sect. A* **2015**, *71*, 3–8. [CrossRef]
61. Sheldrick, G.M. A short history of SHELX. *Acta Crystallogr. Sect. A* **2008**, *64*, 112–122. [CrossRef]

Disclaimer/Publisher's Note: The statements, opinions and data contained in all publications are solely those of the individual author(s) and contributor(s) and not of MDPI and/or the editor(s). MDPI and/or the editor(s) disclaim responsibility for any injury to people or property resulting from any ideas, methods, instructions or products referred to in the content.

Article

Synthesis, Spectroscopic Characterization, and Photophysical Studies of Heteroleptic Silver Complexes Bearing 2,9-Bis(styryl)-1,10-phenanthroline Ligands and Bis[(2-diphenylphosphino)phenyl] Ether

Dimitrios Glykos [1], Athanassios C. Tsipis [1], John C. Plakatouras [1,2] and Gerasimos Malandrinos [1,*]

1. Laboratory of Inorganic Chemistry, Department of Chemistry, University of Ioannina, 451 10 Ioannina, Greece; d.glykos@uoi.gr (D.G.); attsipis@uoi.gr (A.C.T.); iplakatu@uoi.gr (J.C.P.)
2. Institute of Materials Science and Computing, University Research Center of Ioannina, 451 10 Ioannina, Greece
* Correspondence: gmalandr@uoi.gr

Citation: Glykos, D.; Tsipis, A.C.; Plakatouras, J.C.; Malandrinos, G. Synthesis, Spectroscopic Characterization, and Photophysical Studies of Heteroleptic Silver Complexes Bearing 2,9-Bis(styryl)-1,10-phenanthroline Ligands and Bis[(2-diphenylphosphino)phenyl] Ether. *Inorganics* **2024**, *12*, 131. https://doi.org/10.3390/inorganics12050131

Academic Editor: Binbin Chen

Received: 7 April 2024
Revised: 26 April 2024
Accepted: 30 April 2024
Published: 2 May 2024

Copyright: © 2024 by the authors. Licensee MDPI, Basel, Switzerland. This article is an open access article distributed under the terms and conditions of the Creative Commons Attribution (CC BY) license (https://creativecommons.org/licenses/by/4.0/).

Abstract: Three new heteroleptic Ag(I) complexes, labeled as [AgL(POP)]BF$_4$ (**1–3**), were successfully synthesized and comprehensively characterized. Here, L represents 2,9-bis((E)-4-methoxystyryl)-1,10-phenanthroline (**L1**), 2,9-bis((E)-4-methylthiostyryl)-1,10-phenanthroline (**L2**), and 2,9-bis((E)-4-diethylaminostyryl)-1,10-phenanthroline (**L3**), while POP stands for Bis[(2-diphenylphosphino)phenyl] ether. The stability of these compounds in solution was confirmed through multinuclear 1D (^{1}H, ^{13}C, ^{31}P) and 2D NMR (COSY, NOESY, HMBC, HSQC) spectroscopies. Additionally, their molecular structure was elucidated via X-ray crystallography. The photophysical properties of the complexes were assessed both in the solid state and in solution (dichloromethane). Compounds **1–3** demonstrated moderate emissions in solution, with quantum yields ranging from 11–23%. Interestingly, their solid-state luminescent behavior differed. Large bathochromic shifts (42–75 nm) of the emission maxima and a decrease in quantum yields (2.5–9.5%) were evident, possibly due to the presence of excimers. Compound **3** stands out as a rare example of an Ag(I) red-color emitter.

Keywords: silver(I) heteroleptic complexes; X-ray diffraction; luminescence studies; ^{31}P NMR studies

1. Introduction

Luminescent compounds based on transition metals are increasingly garnering attention for various applications, such as organic light-emitting diodes (OLEDs), photocatalysis, and luminescence sensing and imaging [1–3]. The focal point of these research efforts has predominantly centered on late-transition metal-containing materials, specifically those featuring third-row transition metal complexes with d^6 or d^8 electronic configurations, which serve as prominent models. Nonetheless, there exists a developing interest in the exploration of emissive metal complexes with d^{10} configurations due to their abundant availability [4–6].

Group 11 metal(I) complexes capable of emitting light through phosphorescence and/or thermally activated delayed fluorescence (TADF) at room temperature have emerged as promising emitters for OLED devices because they can utilize both triplet (T1) and singlet (S1) excitons. Among them, silver(I) complexes are potential candidates for highly efficient emitters with shorter decay times. However, the inherent redox potential of the silver ion limits metal-to-ligand charge transfer (MLCT) characteristics, often leading to emissions dominated by ligand-centered ππ* transitions. Consequently, luminescent Ag(I) complexes displaying TADF are uncommon [7,8].

Among the most promising emissive silver(I) compounds are heteroleptic [Ag(P^P)(N^N)]$^+$ species, where N^N represents an aromatic diimine type ligand and P^P denotes a sterically demanding bis(phosphino) chelate such as Bis[(2-diphenylphosphino)phenyl] ether (POP).

The coordination chemistry of the large-bite bidentate POP ligand has been subject to an extensive investigation by Van Leeuwen and colleagues, as well as by other researchers [9,10]. This interest stems from its adaptable coordination behavior and its efficacy as a catalyst in various organic transformations. Copper(I)-POP species have undergone extensive scrutiny, whereas analogous silver complexes are less studied. On the other hand, a careful choice of the N^N ligand is also important to optimize the complexes' photophysical characteristics. Rigid structures based on 2,2'-bipyridine or 1,10-phenanthroline cores are commonly employed, while the extent of conjugation and the electronic properties of the N^N ligand are more easily controlled than those of the P^P ligand [6,7,11–13].

Building upon our previous research on the coordination chemistry and luminescent properties of heteroleptic Ir(III), Cu(I), and Ag(I) compounds [14–18], herein, we present our findings on mixed-ligand emissive Ag(I) complexes comprising POP (P^P ligand) and 2,9-Bis(styryl)-1,10-phenanthroline derivatives (N^N ligands). The latter differ only in the nature of the X group located in position 4 of the styryl ring. (X= -methoxy, -methythio, -diethylamino). We believe that it would also be of interest to examine the impact of these electron-donating substituents on the structural, optical, and photophysical properties of the compounds.

2. Results and Discussion

2.1. Synthesis and Characterization

The N^N type ligand 2,9-bis((E)-4-methoxystyryl)-1,10-phenanthroline (**L1**) was synthesized according to a previously described method [19]. The same procedure was followed for the new ligands 2,9-bis((E)-4-methylthioystyryl)-1,10-phenanthroline (**L2**), and 2,9-bis((E)-4-dietlylaminostyryl)-1,10-phenanthroline (**L3**) (experimental details are provided in Supplementary Materials). The synthetic route yielding the heteroleptic silver(I) complexes containing the aforementioned ligands and POP is illustrated in Scheme 1.

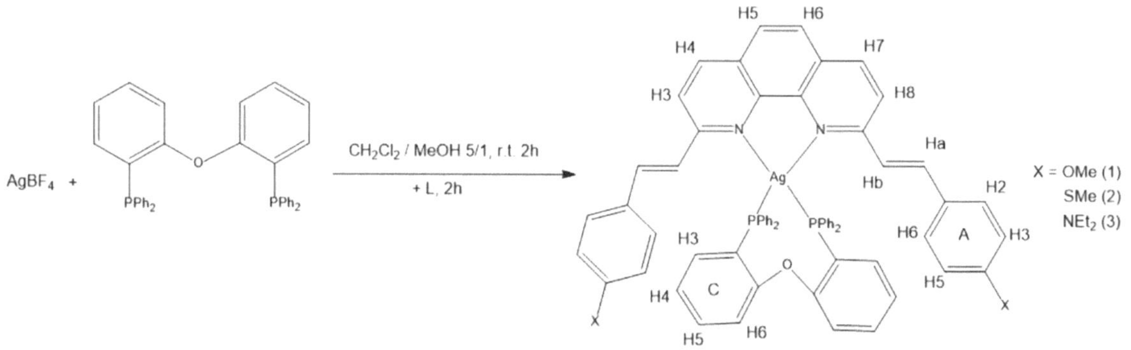

Scheme 1. The synthetic procedure for complexes **1–3**, along with the atom numbering used for NMR assignments.

The procedure involves the reaction of AgBF$_4$ with POP (2 h at room temperature), followed by the addition of **L1–L3** Unfortunately, only complex **2** crystallized as a BF$_4^-$ salt, while crystallization of **1** and **3** was achieved using the same synthetic procedure, but with a different Ag(I) source (AgPF$_6$). However, it should be noted that all the solution work was performed on the [Ag(N^N)(POP)][BF$_4$] compounds. In the ATR-IR spectra (Figures S1–S3), all complexes exhibited absorption bands located in the range of 1575–1460 cm^{-1} and ascribed to ν(C=N) and ν(C=C) (POP and N^N ligands). In addition, the characteristic absorption band of the anion (BF$_4^-$) was observed at approximately 1050 cm^{-1} (ν(B-F)) [20,21].

2.2. Characterization in Solution

The ^{1}H NMR and ^{13}C{^{1}H} NMR spectra of ligands and complexes in CDCl$_3$ are provided in the ESI, Figures S4–S16. The ^{1}H chemical shifts (δ, ppm), as well as Δδ values (Δδ = δ$_{complex}$ − δ$_{ligand}$) derived from the spectra analysis, are listed in Table 1. The ^{31}P{^{1}H} NMR spectra in CDCl$_3$ of all complexes are depicted in Figure 1. In the ^{1}H NMR spectra, 1–3 exhibit sharp and resolved signals, indicating the integrity of the compounds. The data presented in Table 1, and especially the chemical shifts differences induced by complexation (Δδ), confirm the strong interaction of both ligands with Ag(I). The double bonds of the two styryl moieties adopt an E-configuration, inferred by the calculated J-coupling of 16 Hz.

Table 1. ^{1}H-NMR data (δ/ppm) for complexes **1–3**.

H Atoms	1	2	3	Δδ (1) *	Δδ (2) *	Δδ (3) *
H(3)/H(8)	8.24	8.28	8.14	0.31	0.34	0.24
H(4)/H(7)	8.44	8.48	8.32	0.22	0.25	0.16
H(5)/H(6)	7.83	7.86	7.77	0.09	0.11	0.09
H(a)	7.48	7.51	7.39	−0.29	−0.26	−0.33
H(b)	7.42	7.48	7.30	−0.2	−0.21	−0.24
A, H(2,6)	6.85	6.95	6.70	−0.83	−0.69	−0.9
A, H(3,5)	6.65	6.81	6.33	−0.34	−0.51	−0.4
(-OMe) (-CH$_3$)	3.88	-	-	−0.02	-	-
(-SMe) (-CH$_3$)	-	2.56	-	-	0	-
(-NEt$_2$) (-CH$_2$)	-	-	3.42	-	-	−0.03
(-NEt$_2$) (-CH$_3$)	-	-	1.25	-	-	−0.01
POP(C ring) H3	6.73	6.74	6.85	0.01	0.02	0.13
POP(C ring) H4	7.19	7.21	7.02	−0.03	−0.01	−0.20
POP(C ring) H5	7.12	7.12	7.15	0.13	0.13	0.16
POP(C ring) H6	6.91	6.92	6.93	0.07	0.08	0.09

* Δδ = δ$_{complex}$ − δ$_{ligand}$.

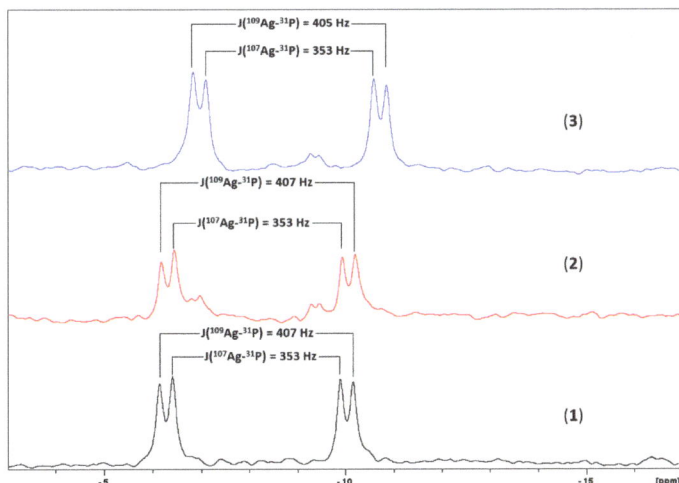

Figure 1. ^{31}P{^{1}H} NMR spectrum of **1–3** in CDCl$_3$ (101.25 MHz, 298 K).

Further insight into the Ag(I) coordination sphere and geometry was obtained through ^{31}P NMR spectroscopy. A well-resolved doublet of doublets appeared for all complexes due to the coupling of ^{107}Ag and ^{109}Ag nuclei with P atoms. This phenomenon has also been observed for similar complexes in the literature. The values of $J(^{107}\text{Ag-}^{31}\text{P})$ and $J(^{109}\text{Ag-}^{31}\text{P})$ shown in Figure 1 suggest that Ag(I) adopts a tetrahedral geometry employing the diimine ligand and the phosphine POP [22,23].

2.3. Optical-Photophysical Properties of Ligands in Solution

The optical and photophysical characteristics of ligand L1 were previously documented in our earlier work (λ_{em} = 433 nm, Φ_{em} = 6% in DCM) [17]. Figure 2 displays the UV-Vis and emission spectra of ligands L2 and L3 in CH_2Cl_2, with corresponding photophysical data outlined in Table 2. Both ligands exhibit absorption bands around 240 nm and within the 330–380 nm wavelength range. However, only L3 displays a noticeable absorption beyond 400 nm. The former is likely attributed to ligand-centered $\pi \to \pi^*$ and $n \to \pi^*$ transitions, while the characteristic band beyond 400 nm is probably associated with ILCT transitions. Ligands L2 and L3 demonstrate luminescence exclusively in solution (dichloromethane, RT). The relative photoluminescent quantum yields (Φ_{em}) calculated for L2 and L3 are 10% and 6%, respectively. The influence of altering the electron-donating substituent at position 4 of the styryl moiety is evident in the position of the emission band. L3, which contains the strongest donor (-NEt$_2$), emits at a significantly lower energy compared to L1 and L2.

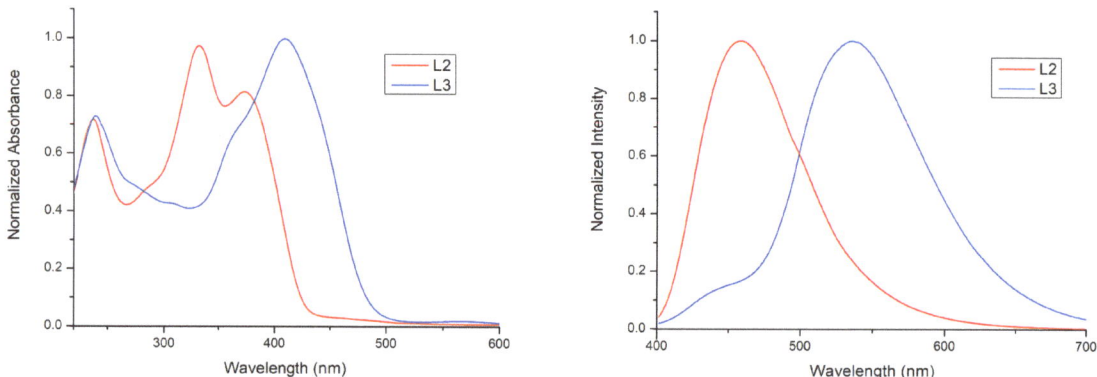

Figure 2. The UV-vis spectrum of ligands L2 and L3 (on the **left**) and their emission spectra in dichloromethane (10^{-5} M, λ_{exc} = 350 nm) (on the **right**).

Table 2. Selected photophysical data for ligands L2 and L3 in both solution and solid states.

Compound	λ_{abs}/nm (ε/M^{-1}cm^{-1}) (CH_2Cl_2)	λ_{em}/nm (CH_2Cl_2)	Φ_{em} (CH_2Cl_2)	λ_{em}/nm (Solid)	Φ_{em} (Solid)
L2	235 (44,000), 332 (60,000), 377 (48,000)	460	10%	-	-
L3	239 (45,454), 360 (40,184), 407 (62,582)	538	6%	-	-

2.4. Optical Properties of Complexes in Solution

The UV-Vis and emission spectra of complexes in CH_2Cl_2 (10^{-5} M) at 298 K are depicted in Figure 3, and the corresponding absorption and photophysical data are summarized in Table 3.

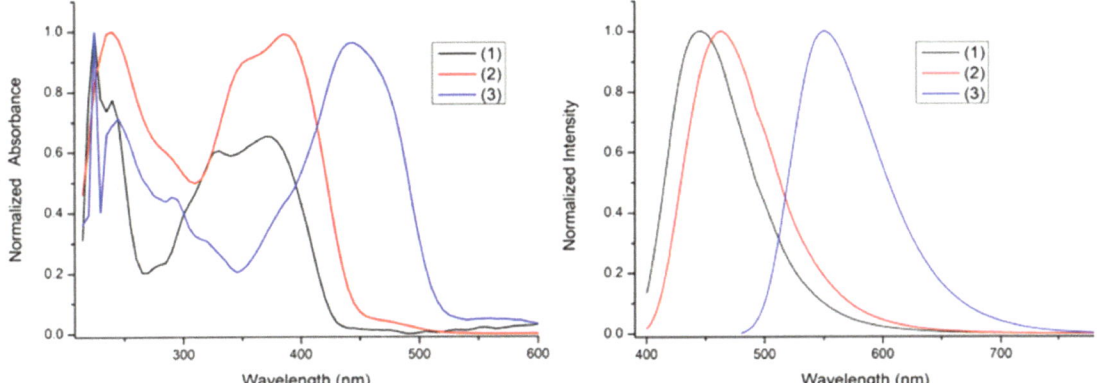

Figure 3. The UV-Vis spectra of complexes **1–3** (on the **left**) and the emission spectra (λ_{exc} = 350 nm) in dichloromethane (298 K, 10^{-5} M) (on the **right**).

Table 3. Selected photophysical data for compounds **1–3** both in solution and solid state.

Compound	λ_{abs}/nm (ε/M^{-1}cm^{-1}) (CH$_2$Cl$_2$)	λ_{em}/nm (CH$_2$Cl$_2$)	Φ_{em} (CH$_2$Cl$_2$)	λ_{em}/nm (Solid)	Φ_{em} (Solid)
1	230 (96,153), 335 (62,500), 385 (65,384)	445	11%	487	9.5%
2	235 (63,532), 350 (55,908), 390 (60,991)	465	20%	534	5.6%
3	245 (56,112), 445 (80,160)	550	23%	625	2.5%

The absorption characteristics of complexes **1** and **2** are nearly identical across the entire recorded wavelength region. Both exhibit absorption bands at 240 nm and within the 330–380 nm range, with no discernible absorption observed beyond 450 nm. The former is attributed to ligand-centered (LC) $\pi \rightarrow \pi^*$ transitions, while the latter is expected to have a combined character of ligand-to-ligand charge transfer (LLCT) and metal-to-ligand charge transfer (MLCT). This behavior is commonly observed in [Ag (N^N)(P^P)] type complexes [20,21]. In contrast, the absorption spectrum of complex **3** is notably different. A strong band centered at λ = 450 nm is evident, possibly attributed to MLCT and ILCT (intraligand) transitions.

Complexes **1** and **2** display nearly identical emission wavelengths (λ_{em}). A slight red shift in the emission maxima of about 20 nm can be observed when transitioning from the –OMe to the –SMe derivative. However, the most notable difference lies in the quantum yield (Φ_{em}). Complex **2** shows an almost twofold increase in Φ_{em} (20%) compared to **1** (11%), suggesting a more efficient radiative relaxation of the excited state for the former.

Complex **3** displays a distinct emission spectrum profile. The strong electron-donating properties of diimine **L3**, attributed to the presence of the (-NEt$_2$) group, effectively destabilize Ag(I) 4d orbitals, resulting in a 100 nm red-shift of the emission maximum compared to **1** and **2**, with a Φ_{em} = 22%. Solution luminescence studies for mononuclear silver(I) complexes are limited, as most are only investigated in the solid state [20–25]. In this context, compounds **1–3** demonstrate superior photoluminescence characteristics in solution, which are easily tunable through the replacement of a single group present on the diimine ligand.

2.5. Luminescent Behavior in Solid State

The diffuse reflectance (DRS) and emission spectra (λ_{exc} = 350 nm) of the complexes obtained in the solid state are illustrated in Figure 4, with corresponding data listed in Table 3. The emission spectra of 1–3 exhibit broad and unstructured profiles, consistent with the expectations for luminescent Ag[(N^N)(P^P)]-type complexes [26–29]. The significant bathochromic shifts observed for λ_{em} (42–75 nm) when transitioning from solution to the solid state, particularly for complexes **2** and **3**, may suggest the formation of excimers. As we will discuss later while examining the crystal structures of the compounds, the presence of C-H···π intermolecular interactions, with strength following the order **3 > 2 > 1**, may account for this phenomenon [30,31]. Simultaneously, the Φ_{em} values for complexes **2** and **3** drop significantly compared to their solution states. This observation suggests that the excited dimers may undergo substantial energy loss, likely due to geometrical reorganization and/or Jahn–Teller distortion, in the excited state. Finally, it is noteworthy that complex **3**, emitting at λ_{em} = 625 nm, represents a rare instance of an Ag(I) mononuclear red emitter [32–35].

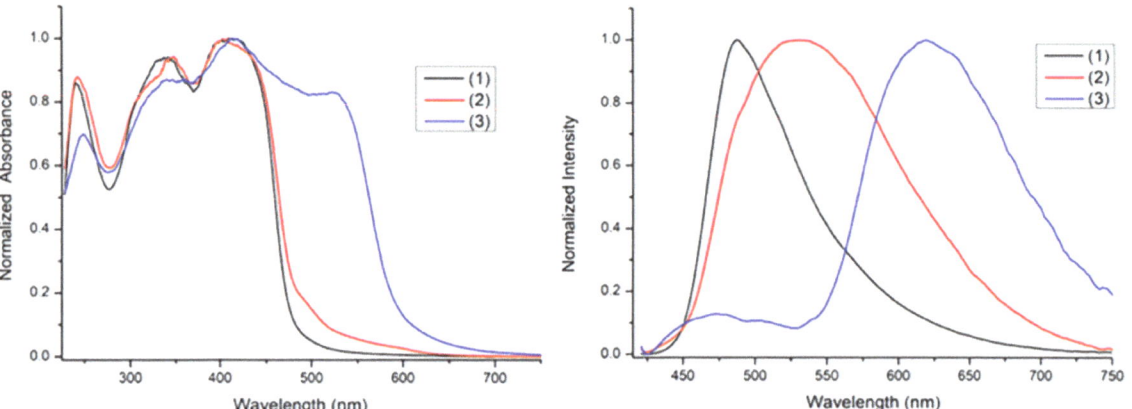

Figure 4. The DRS spectra of complexes **1–3** (on the **left**) and the emission spectra (λ_{exc} = 350 nm) in solid state (on the **right**).

2.6. Single-Crystal Structure Analysis

High-quality X-ray crystals were obtained by introducing diethyl ether vapor into a solution of complexes. In instances where crystal growth posed challenges, various solvents and crystallization techniques were employed.

Compound **1** (PF_6^-) crystallizes in the triclinic space group P-1. A visual representation of the cation's structure, as well as a section of the packing, is presented in Figure 5. Important bond distances (in Å) and angles (in degrees) within the coordination sphere of Ag(I) in the cation are listed in Table 4.

Table 4. Selected structural characteristics of **1** (PF_6^-).

Bond Distances	(Å)	Bond Angles	(°)
Ag(1)-N(1)	2.350(3)	N(1)-Ag(1)-N(2)	72.11(12)
Ag(1)-N(2)	2.356(4)	N(1)-Ag(1)-P(2)	127.15(9)
Ag(1)-P(2)	2.4104(12)	N(2)-Ag(1)-P(2)	133.36(9)
Ag(1)-P(1)	2.5661(12)	N(1)-Ag(1)-P(1)	97.05(9)
		N(2)-Ag(1)-P(1)	96.20(9)
		P(2)-Ag(1)-P(1)	118.64(4)

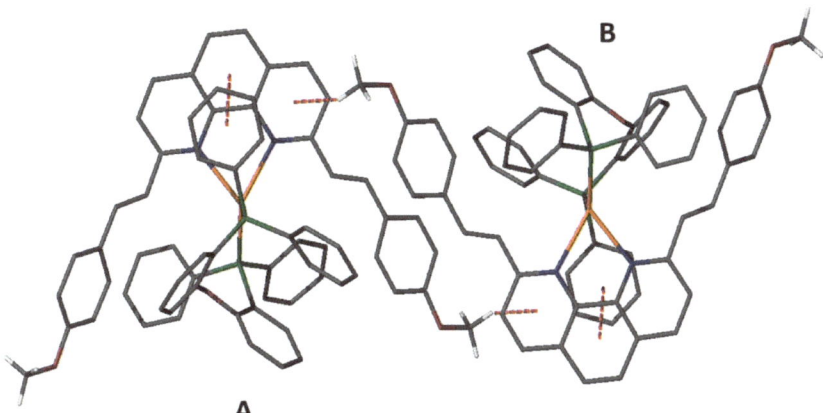

Figure 5. A section of the packing in the crystal structure of complex **1**, illustrating the intermolecular CH...π and π...π stacking interactions (red dotted lines). Symmetry operations to generate equivalent atoms: (**A**), x, y, z; (**B**), 2 − x, 1 − y, 1 − z. For clarity, the counter-anion and aromatic hydrogen atoms are omitted. Atom labeling: Ag (orange), N (blue), P (green), and O (red).

Compound **2** crystallizes in the monoclinic space group P21/c. The asymmetric unit comprises the cation [AgL(POP)]$^+$ and its corresponding counter-anion, BF$_4^-$. A visual representation of the cation's structure, as well as a section of the packing, can be found in Figure 6, while the structural characteristics are detailed in Table 5.

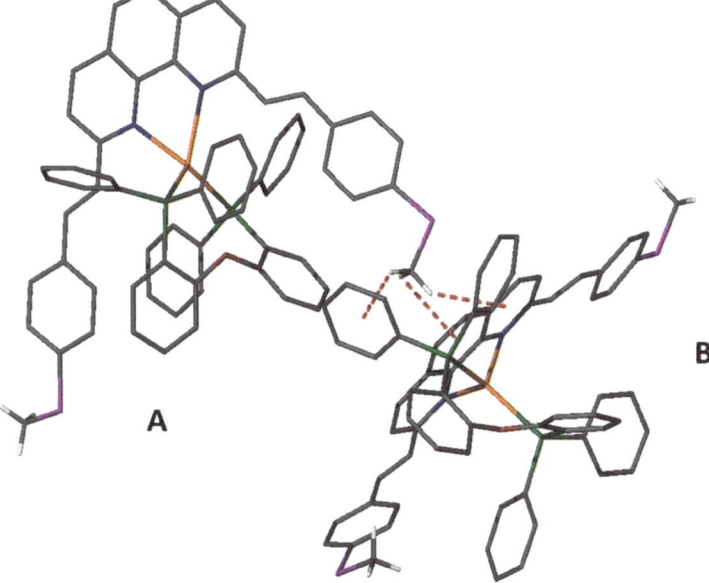

Figure 6. A section of the packing in the crystal structure of complex **2**, illustrating the intermolecular CH...π stacking interactions (red dotted lines). Symmetry operations to generate equivalent atoms: (**A**), x, y, z; (**B**), 1 − x, 1/2 + y, 3/2 − z. For clarity, the counter-anion and aromatic hydrogen atoms are omitted. Atom labeling: Ag (orange), N (blue), P (green), O (red), and S (purple).

Table 5. Selected structural characteristics of **2** (BF$_4^-$).

Bond Distances	(Å)	Bond Angles	(°)
Ag(1)-N(1)	2.385(3)	N(1)-Ag(1)-N(2)	70.43(11)
Ag(1)-N(2)	2.392(3)	N(1)-Ag(1)-P(2)	114.60(9)
Ag(1)-P(2)	2.5240(12)	N(2)-Ag(1)-P(2)	118.77(8)
Ag(1)-P(1)	2.5245(12)	N(1)-Ag(1)-P(1)	117.89(8)
		N(2)-Ag(1)-P(1)	110.36(9)
		P(2)-Ag(1)-P(1)	116.67(4)

Compound **3** (PF$_6^-$) crystallizes in the trigonal space group P 31 2 1. Figure 7 illustrates the cation geometry and a section of the packing. Values for the most significant bonds and angles are summarized in Table 6.

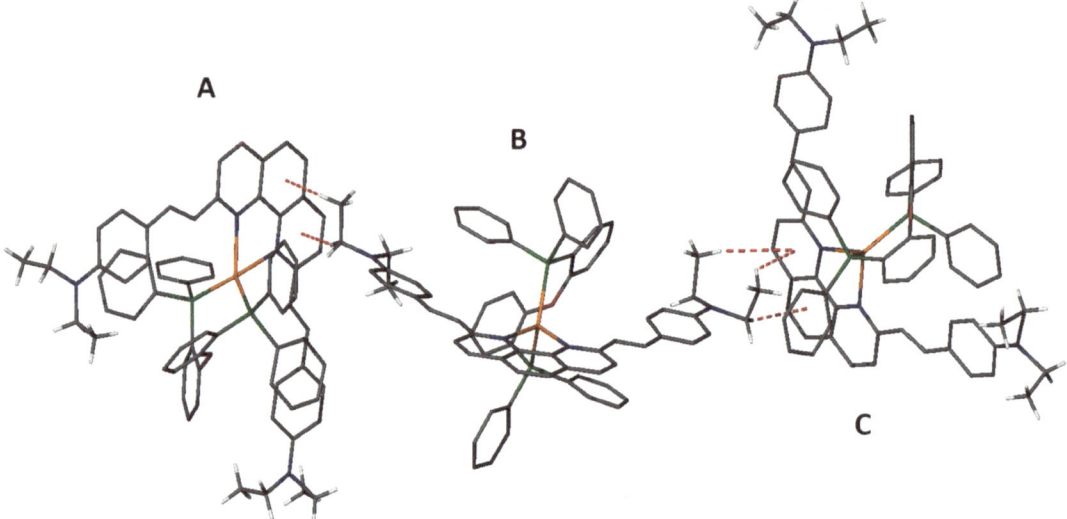

Figure 7. A section of the packing in the crystal structure of complex **3**, illustrating the intermolecular CH...π stacking interactions (red dotted lines). Symmetry operations to generate equivalent atoms: (**A**), 1 − y, 1 + x − y, 1/3 + z; (**B**), x, y, z; and (**C**), 1 − x + y, 1 − x, z − 1/3. For clarity, the counter-anion and aromatic hydrogen atoms are omitted. Atom labeling: Ag (orange), N (blue), P (green), and O (red).

Table 6. Selected crystallographic data of **3** (PF$_6^-$).

Bond Distances	(Å)	Bond Angles	(°)
Ag(1)-N(1)	2.362(4)	N(1)-Ag(1)-N(2)	71.13(12)
Ag(1)-N(2)	2.377(4)	N(1)-Ag(1)-P(2)	114.57(9)
Ag(1)-P(2)	2.5117(14)	N(2)-Ag(1)-P(2)	118.83(9)
Ag(1)-P(1)	2.5144(13)	N(1)-Ag(1)-P(1)	116.75(9)
		N(2)-Ag(1)-P(1)	113.40(9)
		P(2)-Ag(1)-P(1)	115.11(5)

X-ray structural analysis revealed the formation of mononuclear heteroleptic complexes, where the silver(I) cation adopts a distorted tetrahedral environment with N2P2 coordination. Both the phenanthroline and diphosphine moieties act as chelating ligands. Notably, the POP ligand binds to the metal solely through its pair of P donor atoms, while the ether O atom remains at a nonbonding distance from the Ag(I) center (with Ag(1)−O(1) ranging from 3.173 to 3.199 Å). The bond distances of Ag–P and Ag–N, as well as the chelating angles N–Ag–N and P–Ag–P, exhibit typical values across all complexes [32–35].

Interestingly, π–π stacking interactions involving the phenathroline cores in the crystal lattice were not observed. Instead, C-H···π intermolecular interactions between –OMe, -SMe, -NEt$_2$, and adjacent aromatic (phen/and or phenyl) rings predominate. It is noteworthy that the strength of these interactions follows the order –OMe (4.22 Å) < -SMe (3.30 Å) < -NEt$_2$ (3.03 Å) (the reported values refer to the C-H···phen core distances).

2.7. Computational Study

To obtain further insight into the photophysical properties of the new Ag(I) complexes under study, we employed TDDFT electronic structure calculations in order to simulate and assign their UV-Vis absorption spectra. The simulated absorption spectra of **1–3** in DCM are depicted in Figure 8.

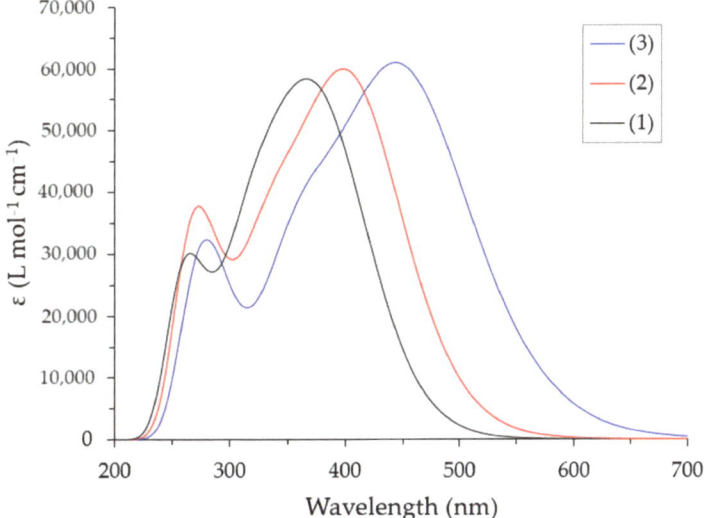

Figure 8. Absorption spectra of **1–3** calculated at the PBE0/LANL2TZ(Ag)U6-31G(d,p) level of theory in DCM solvent.

Inspection of Figure 8 reveals that the simulated absorption spectra reproduce the respective experimentally recorded spectra of **1–3**. Accordingly, the simulated spectra of **1** and **2** exhibit a low-energy band peaking at 370 and 400 nm, respectively, in excellent agreement with the experimental spectra showing peaks at 372 and 390 nm, respectively. In addition, the simulated spectra of **1** and **2** show high-energy bands in the region of 250–280 nm, in line with the respective experimental findings. On the other hand, the simulated absorption spectrum of **3** exhibits a low-energy band at 447 nm, in excellent agreement with the experiment, as well as a high-energy band peaking around 280 nm, in line with the experimental spectrum of **3**. It should be noticed, however, that, although the simulated spectrum of **3** is qualitatively similar to those found for **1** and **2**, there is a striking difference with respect to its low-energy band. Thus, the latter appears in the visible, being red-shifted by about 30–50 nm, as compared to the respective low-energy bands of **1** and **2**, which in contrast appears in the ultraviolet. Let us now assign the UV-Vis spectra in terms of electronic excitations between MOs. In Table 7, the most intense electronic transitions corresponding to the two bands appearing in the absorption spectra of **1–3** in DCM are given.

Table 7. Principal singlet–singlet electronic transitions in the simulated absorption spectra for complexes **1**–**3** calculated in DCM solvent at the PBE0/LANL2TZ(Ag)U6-31G(d,p) level of theory [a].

Excitation (% Composition)	λ (nm)	f	Assignment
	1		
H → L (90%)	397	0.723	MLCT/IL/LL'CT
H-1 → L (62%), H → L + 1 (27%), H-2 → L + 1 (8%)	356	0.352	IL/LL'CT
H → L + 2 (77%), H-2 → L + 1 (5%), H-2 → L + 2 (5%)	324	0.482	MLCT/IL/LL'CT
H → L + 4 (38%), H-2 → L + 3 (23%), H → L + 3 (11%), H-2 → L + 4 (10%)	285	0.056	IL/LL'CT
H-12 → L (16%), H-5 → L + 1 (16%), H-9 → L + 1 (13%)	263	0.058	IL/LL'CT
H-2 → L + 6 (35%), H → L + 6 (17%), H → L + 9 (10%)	255	0.087	IL/LL'CT
	2		
H → L (94%)	420	0.846	IL
H-1 → L (90%), H → L + 1 (6%)	400	0.575	IL
H-1 → L + 1 (92%)	358	0.136	IL
H-6 → L (55%), H-4 → L + 1 (16%)	299	0.093	IL
H-6 → L + 1 (38%), H-8 → L (9%), H-3 → L + 2 (9%)	278	0.078	IL
H-4 → L + 2 (34%), H-12 → L (17%)	266	0.089	IL
	3		
H → L (90%), H-1 → L (6%)	475	0.827	IL
H-1 → L (90%), H → L (7%)	431	0.658	IL
H → L + 2 (95%)	372	0.540	IL
H-4 → L (83%)	306	0.078	IL
H → L + 9 (48%), H → L + 11 (12%)	285	0.160	IL
H-11 → L (28%), H-2 → L + 6 (26%), H-1 → L + 11 (19%)	264	0.098	IL

[a] H and L stand for HOMO and LUMO, respectively.

Perusal of Table 7 reveals that the low-energy bands in the absorption spectra of **1**–**3** in DCM arise mainly from three electronic transitions. Thus, the band peaking around 350 nm in the absorption spectrum of **1** in DCM arises mainly from electronic transitions appearing at 397, 356, and 324 nm. The former is the most intense, and this is due to a HOMO to LUMO excitation. The other two electronic transitions are due to a combination of a multitude of electronic excitations involving the HOMO-1, HOMO, LUMO, LUMO + 1, and LUMO + 2 MOs. In Figure 9, the 3D isocontour surfaces of the MOs involved in the electronic excitations relevant to the electronic transitions related to the low-energy band around 350 nm are depicted in the simulated absorption of **1** in DCM. The MOs depicted in Figure 9 are mainly located on the ligands, with the exception of HOMO, which also exhibits a small localization on the metal. Therefore, the electronic transitions at 397 and 324 nm could be assigned as of Metal to Ligand Charge Transfer (MLCT), Intraligand (IL) and Ligand to Ligand Charge Trans-fer (LL'CT) mixed character (MLCT/IL/LL'CT). On the other hand, the electronic transition at 356 nm could be assigned as of IL/LL'CT character. Consequently, the band around 350 nm is assigned as of MLCT/IL/LL'CT character.

Next, the high-energy band around 250 nm, appearing in the simulated absorption spectrum of **1** in DCM, arises mainly from three electronic transitions at 285, 263, and 255 nm (Table 7). These electronic transitions are due to a multitude of electronic excitations which exhibit IL/LL'CT characters, and, accordingly, the band at 250 nm could assigned the same way (IL/LL'CT).

The low-energy band appearing around 400 nm in the simulated absorption spectrum of **2** in DCM arises mainly from three electronic transitions at 420, 400, and 358 nm. The former is almost solely due a HOMO to LUMO electronic excitation, which occurs on the diimine ligand (see Figure S17 of the Supporting Information). Thus, the electronic excitation at 420 nm is characterized as IL.

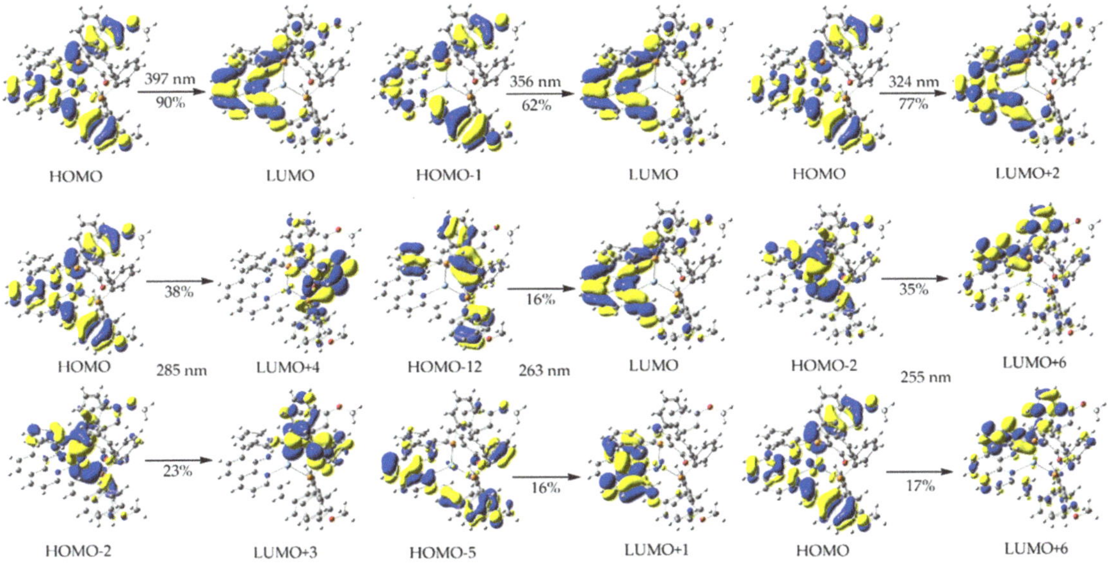

Figure 9. Three-dimensional contour plots of the MOs involved in the electronic excitations giving rise to the two bands appearing in the simulated absorption spectra of **1**, calculated at the PBE0/LANL2TZ(Ag)∪6-31G(d,p) level of theory in DCM solvent.

The electronic transition at 400 nm is basically due to a HOMO-1 to LUMO excitation, and is also characterized as IL (Figure S17). The same also holds true for the electronic transition at 358 nm, which is due to a HOMO-1 to LUMO + 1 excitation, both of which are located on the diimine ligand.

Next, the high-energy band appearing around 260 nm arises mainly from three electronic transitions at 299, 278, and 266 nm (Table 7). Based upon the shapes of the Mos, all these transitions exhibit IL character; therefore, the band at 260 nm is assigned as IL as well.

The absorption spectrum of **3** in DCM, as mentioned earlier (vide supra), is distinct compared to those found for the other two complexes under study, namely, **1** and **2**. Accordingly, the low-energy band appears in the visible around 450 nm. This band arises from three electronic transitions at 471, 431, and 372 nm (Table 7). These electronic transitions are due to excitations involving HOMO-1, HOMO, LUMO, and LUMO + 2, which are located on the diimine ligands. Thus, the low-energy band at 450 nm could be assigned as IL (Figure S18).

Finally, the high-energy band around 270 nm appearing in the simulated absorption spectrum of **3** in DCM arises mainly from three electronic transitions at 306, 285, and 264 nm. Based upon the shapes of the MOs involved in these electronic transitions, the band at 270 nm could be assigned as IL.

3. Materials and Methods

3.1. Materials

All solvents were of analytical grade and used without further processing. $AgBF_4$, $AgPF_6$, POP, 4-methylthiobenzaldehyde, 4-diethylaminobenzaldehyde, and 4-anisaldehyde were purchased from Sigma-Aldrich (Burlington, MA, USA), while neocuproine was obtained from TCI Chemicals (Tokyo, Japan).

3.2. Methods

The instruments and procedures used for acquiring ^{1}H, ^{13}C{^{1}H}, and ^{31}P {^{1}H} NMR spectra, ATR-IR spectra, and emission spectra (in both solution and solid state), as well as DRS and UV-Vis absorption spectra, were similar to those described previously [14–18]. The luminescence quantum yields of complexes **1**, **2**, and **3** in CH_2Cl_2 solutions were determined and calculated at room temperature using $[Ru(bpy)_3]Cl_2$ in water as the reference (Φ_{em} = 0.04) [36].

3.3. Crystal Structure Determination

Single crystals of complexes **1–3** were selected from crystallization solutions and mounted on a Bruker D8 Quest Eco diffractometer. Measurements were conducted using graphite-monochromatic Mo-Kα radiation (λ = 0.71073 Å) and a Photon II detector. The structure was solved using direct methods, and the ShelXle interface enabled full-matrix least-squares methodology to be used on F^2. The SQUEEZE procedure of the PLATON program was utilized to remove disordered counter-anion molecules. Non-hydrogen atoms of the complexes were refined by anisotropic thermal parameters, while hydrogen atoms were refined by isotropic thermal parameters and constrained to ride on their parent atoms. X-Seed software (Version 4.10) was employed to generate molecular graphics [37–42].

In complex **3** (PF_6^-), one ethyl group was found to be disordered, possibly due to its lack of interaction with other lattice constituents. Unfortunately, the counter-anion in this compound was severely disordered and could not be adequately modeled. As a final solution, it was eliminated using the SQUEEZE routine from the PLATON program.

Crystal data for 1 (PF_6^-): $C_{132}H_{103}Ag_2F_{12}N_4O_6P_6$, M = 2470.75, yellow polyhedral, 0.50 × 0.32 × 0.28 mm^3, triclinic, space group P-1 (No. 2), a = 17.5294(8), b = 18.6280(9), c = 19.9345(9) Å, α = 70.918(3), β = 67.775(2), γ = 82.362(3)°, V = 5694.3(5) Å^3, Z = 2, D_c = 1.441 g cm^{-3}, F_{000} = 2526, Bruker D8 Eco with PHOTON II detector, MoKα radiation, λ = 0.71073 Å, T = 296(2) K, $2\theta_{max}$ = 50.0°, 230,911 reflections collected, 17,980 unique (R_{int} = 0.0793). Final $GooF$ = 1.081, $R1$ = 0.0535, $wR2$ = 0.1339, R indices based on 13,600 reflections with $I > 2\sigma(I)$ (refinement on F^2), 1462 parameters, 0 restraints. Lp and absorption corrections applied, μ = 0.508 mm^{-1}.

Crystal data for 2 (BF_4^-): $C_{66}H_{52}AgBF_4N_2OP_2S_2$, M = 1209.83, yellow needle, 0.300 × 0.080 × 0.060 mm^3, monoclinic, space group $P2_1/c$ (No. 14), a = 14.2376(12), b = 20.6038(17), c = 20.3351(16) Å, β = 104.476(3)°, V = 5775.9(8) Å^3, Z = 4, D_c = 1.391 g cm^{-3}, F_{000} = 2480, Bruker D8 Eco with PHOTON II detector, MoKα radiation, λ = 0.71073 Å, T = 296(2) K, $2\theta_{max}$ = 50.0°, 185,416 reflections collected, 10,152 unique (R_{int} = 0.1269). Final $GooF$ = 1.171, $R1$ = 0.0649, $wR2$ = 0.1077, R indices based on 7846 reflections with $I > 2\sigma(I)$ (refinement on F^2), 714 parameters, 0 restraints. Lp and absorption corrections applied, μ = 0.535 mm^{-1}.

Crystal data for 3 (PF_6^-): $C_{72}H_{66}AgN_4OP_2$, M = 1173.10, yellow block, 0.600 × 0.500 × 0.400 mm^3, trigonal, space group $P3_121$ (No. 152), a = 19.328(4), b = 19.328(4), c = 32.247(6) Å, V = 10,433(5) Å^3, Z = 6, D_c = 1.120 g cm^{-3}, F_{000} = 3666, Bruker D8 Eco with PHOTON II detector, MoKα radiation, λ = 0.71073 Å, T = 293(2) K, $2\theta_{max}$ = 50.1°, 127,376 reflections collected, 12,286 unique (R_{int} = 0.0360). Final $GooF$ = 1.079, $R1$ = 0.0366, $wR2$ = 0.0804, R indices based on 11,128 reflections with $I > 2\sigma(I)$ (refinement on F^2), 743 parameters, 64 restraints. Lp and absorption corrections applied, μ = 0.377 mm^{-1}.

CCDC2344379, CCDC2344380 and CCDC2344381 contain the supplementary crystallographic data for this paper. These data can be obtained free of charge via https://www.ccdc.cam.ac.uk/structures/? (accessed on 29 March 2024).

3.4. Computational Details

The TDDFT calculations were performed employing the Gaussian16W software (Version C.02) [43] using the 1997 hybrid functional of Perdew, Burke, and Ernzerhof [44–49]. This functional uses 25% exchange and 75% weighting correlation and is denoted as PBE0. The LANL2TZ basis set was used for the Ag atoms, while for non-metal atoms, we employed the 6-31G(d,p) basis set. The TDDFT calculations were performed for the S_0 ground state using the structures obtained from X-ray analysis and taking into account 50 excited states. Solvent effects were calculated using the Polarizable Continuum Model (PCM) with the integral equation formalism variant (IEF-PCM), which is the default method of G16W (self-consistent reaction field (SCRF)) [50], while Dichloromethane (DCM) was used as the solvent.

3.5. Synthesis of Complexes

[Ag(**L1**)(POP)][BF$_4$] (**1**)

For this step, 0.1 mmol of POP was added to a solution containing 0.1 mmol of AgBF$_4$ in CH$_2$Cl$_2$/MeOH (5:1 v/v) at room temperature. The mixture was stirred for 2 h. Following this, **L1** was introduced, and stirring continued for an additional 2 h. The solvents were then evaporated under vacuum. The resulting yellow solid residue was isolated and dried, yielding 78%.

C$_{66}$H$_{52}$AgBF$_4$N$_2$O$_3$P$_2$: ^{1}H NMR (500 MHz, CDCl$_3$) (ppm): 8.44 (d, J = 8.5 Hz, 2H); 8.24 (d, J = 8.6 Hz, 2H); 7.84 (s, 2H); 7.49 (d, J = 16.3 Hz, 4H); 7.42 (d, J = 16.3 Hz, 4H); 7.05–7.23 (m, 18H); 6.92 (t, J = 7.5 Hz, 8H); 6.85 (d, J = 8.6 Hz, 4H); 6.74 (d, J = 7.9 Hz, 2H); 6.65 (d, J = 8.6 Hz, 4H); 3.89 (s, 6H). ^{13}C NMR (125 MHz, CDCl$_3$) (ppm): 160.5; 157.9; 155.7; 143.1; 138.3; 137; 134; 132.8; 131.9; 130.1; 129; 128.5; 127.9; 127.5; 126.3; 124.87; 120.2; 119.3; 114; 55.4. ^{31}P NMR (101.25 MHz), CDCl3) (ppm): −8.21 (dd, $J(^{109}$Ag-^{31}P) = 407 Hz, $J(^{107}$Ag-^{31}P) = 353 Hz).

HR ESI-MS: m/z = 1091.2487 for [Ag(**L1**)POP)]$^+$ (Figure S19).

[Ag(**L2**)(POP)][BF$_4$] (**2**)

The synthetic procedure for compound **1** was followed, employing **L2** instead (orange solid, yield 70%).

C$_{66}$H$_{52}$AgBF$_4$N$_2$O$_1$S$_2$P$_2$: ^{1}H NMR (500 MHz, CDCl$_3$) (ppm): 8.48 (d, J = 8.5 Hz, 2H); 8.29 (d, J = 8.6 Hz, 2H); 7.86 (s, 2H); 7.52 (d, J = 16.3 Hz, 4H); 7.48 (d, J = 16.3 Hz, 4H); 7.02–7.25 (m, 19H); 6.96 (d, J = 8.3 Hz, 4H); 6.91 (t, 8H); 6.82 (d, J = 8.3 Hz, 4H); 6.73 (m, 2H); 2.56 (s, 6H). ^{13}C NMR (125 MHz, CDCl$_3$) (ppm): 157.9; 155.3; 143.1; 140.2; 138.5; 136.8; 134.1; 133.7; 132.8; 131.9; 130.2; 129; 128.8; 127.8; 126.5; 125.8; 124.8; 120.52; 119.2; 15.4. ^{31}P NMR (101.25 MHz), CDCl3) (ppm): −8.20 (dd, $J(^{109}$Ag-^{31}P) = 407 Hz, $J(^{107}$Ag-^{31}P) = 353 Hz).

HR ESI-MS: m/z = 1121.2020 for [Ag(**L2**)(POP)]$^+$ (Figure S20).

[Ag(**L3**)(POP)][BF$_4$] (**3**)

The synthetic procedure for compound **1** was followed, employing **L3** instead (red solid, yield 68%).

C$_{72}$H$_{66}$AgBF$_4$N$_4$OP$_2$: ^{1}H NMR (500 MHz, CDCl$_3$) (ppm): 8.32 (d, J = 8.6 Hz, 2H); 8.14 (d, J = 8.6 Hz, 2H); 7.77 (s, 2H); 7.40 (d, J = 16.1 Hz, 4H); 7.31 (d, J = 16.3 Hz, 4H); 6.88–7.20 (m, 28H); 6.71 (d, J = 8.6 Hz, 4H); 6.32 (d, J = 7.9 Hz, 4H); 3.43 (q, J = 7.0 Hz, 8H); 1.25 (t, J = 7.1 Hz, 12H). ^{13}C NMR (125 MHz, CDCl$_3$) (ppm): 157.76; 156.26; 148.44; 143.25; 137.6; 134.2; 132.9; 131.8; 130; 129.2; 128.5; 128; 125.6; 124.85; 124.4; 122.2; 119.6; 118.9; 111.18; 44.6; 12.6. ^{31}P NMR (101.25 MHz), CDCl3) (ppm): −8.63 (dd, $J(^{109}$Ag-^{31}P) = 405 Hz, $J(^{107}$Ag-^{31}P) = 353 Hz).

HR ESI-MS: m/z = 1173.3764 for [Ag(**L3**)(POP)]$^+$ (Figure S21).

4. Conclusions

In summary, we successfully synthesized and characterized three new Ag(I) heteroleptic complexes of the Ag(N^N)(P^P) type, which contain 2,9-Bis(styryl)-1,10- phenanthroline ligands and the diphosphine POP. X-ray crystallography revealed that Ag(I) adopts a distorted tetrahedral geometry formed by two chelating nitrogen (N^N) and two phosphorus (P^P) atoms. This structural arrangement persists in solution, as indicated by the NMR data.

Compounds **1**–**3** exhibit superior photoluminescence properties in solution compared to the majority of **the** reported examples. These properties are easily modifiable by replacing a single group on the diimine ligand. However, a different luminescent behavior was observed in the solid state, characterized by large bathochromic shifts and poor emissive properties, possibly due to the formation of excimers. Notably, compound **3** is a rare example of an Ag(I) red emitter in the solid state.

We intend to extend our studies towards optimizing the photophysical properties of similar compounds and fully elucidating the emission mechanism.

Supplementary Materials: The following supporting information can be downloaded at: https://www.mdpi.com/article/10.3390/inorganics12050131/s1, Synthesis and NMR-MS data for ligands L2 and L3. Figures S1–S3: ATR-IR spectra of complexes 1–3, respectively; Figure S4: ^{1}H-NMR spectrum of L2 (500 MHz, CDCl$_3$, 298 K); Figure S5: ^{13}C-NMR spectrum of L2 (500 MHz, CDCl$_3$, 298 K); Figure S6: ^{1}H-NMR spectrum of L3 (500 MHz, CDCl$_3$, 298 K); Figure S7: ^{13}C-NMR spectrum of L3 (500 MHz, CDCl$_3$, 298 K); Figure S8: ^{1}H-NMR spectrum of 1 (500 MHz, CDCl$_3$, 298 K); Figure S9: ^{13}C-NMR spectrum of 1 (500 MHz, CDCl$_3$, 298 K); Figure S10: ^{1}H-^{1}H COSY NMR spectrum of 1 (500 MHz, CDCl$_3$, 298 K); Figure S11: ^{1}H-NMR spectrum of 2 (500 MHz, CDCl$_3$, 298 K); Figure S12: ^{13}C-NMR spectrum of 2 (500 MHz, CDCl$_3$, 298 K); Figure S13: ^{1}H-^{1}H COSY NMR spectrum of 2 (500 MHz, CDCl$_3$, 298 K); Figure S14: ^{1}H-NMR spectrum of 3 (500 MHz, CDCl$_3$, 298 K); Figure S15: ^{13}C-NMR spectrum of 3 (500 MHz, CDCl$_3$, 298 K); Figure S16: ^{1}H-^{1}H COSY NMR spectrum of 3 (500 MHz, CDCl$_3$, 298 K); Figure S17: 3D contour plots of the MOs involved in the electronic excitations giving rise to the two bands appearing in the simulated absorption spectra of **2** calculated at the PBE0/LANL2TZ(Ag)U6-31G(d,p) level of theory in DCM solvent; Figure S18: 3D contour plots of the MOs involved in the electronic excitations giving rise to the two bands appearing in the simulated absorption spectra of **3** calculated at the PBE0/LANL2TZ(Ag)U6-31G(d,p) level of theory in DCM solvent; Figure S19: HR-ESI-MS spectrum of the fragment [AgL]$^+$ **1** (top) and theoretical spectrum; Figure S20: HR-ESI-MS spectrum of the fragment [AgL]$^+$ **2** (top) and theoretical spectrum; Figure S21: HR-ESI-MS spectrum of the fragment [AgL]$^+$ **3** (top) and theoretical spectrum; Figure S22: HR-ESI-MS spectrum of the fragment [L2 + H$^+$]$^+$ (top) and theoretical spectrum; Figure S23: HR-ESI-MS spectrum of the fragment [L3 + H$^+$]$^+$ (top) and theoretical spectrum; File S1: gm7_23_a_sq (*.cif file); File S2: gm7_23_a_sq (checkcif, pdf file); File S3: GM16_22B_0m (*.cif file); File S4: GM16_22B_0m (checkcif, pdf file); File S5: GM31_23_0m_a (*.cif file); File S5: GM31_23_0m_a (checkcif, pdf file).

Author Contributions: Conceptualization: G.M.; supervision: G.M.; formal analysis: D.G. and J.C.P.; investigation: D.G.; computational study: A.C.T.; X-ray crystallography: J.C.P. and D.G.; writing—original draft: D.G.; writing—review and editing: G.M. and J.C.P. All authors have read and agreed to the published version of the manuscript.

Funding: This research received no external funding.

Data Availability Statement: The information provided in this research is accessible in both the article and its Supplementary Materials. CCDC2344379, CCDC2344380, and CCDC2344381 contain the supplementary crystallographic data for this paper. These data can be obtained free of charge via https://www.ccdc.cam.ac.uk/structures/ (accessed on 29 March 2024). All other data in this study can be found in the Supplementary Materials.

Acknowledgments: The authors would like to thank the Network of Research Supporting Laboratories at the University of Ioannina for providing access to the use of NMR, ESI-MS, and X-ray diffraction facilities.

Conflicts of Interest: The authors declare no conflicts of interest.

References

1. Fernández-Moreira, V.; Thorp-Greenwood, F.L.; Coogan, M.P. Application of D6 Transition Metal Complexes in Fluorescence Cell Imaging. *Chem. Commun.* **2010**, *46*, 186–202. [CrossRef]
2. Yam, V.W.W.; Wong, K.M.C. Luminescent Metal Complexes of D6, D8 and d 10 Transition Metal Centres. *Chem. Commun.* **2011**, *47*, 11579–11592. [CrossRef]
3. Lo, K.K.-W.; Choi, A.W.; Law, W.H.-T. Applications of Luminescent Inorganic and Organometallic Transition Metal Complexes as Biomolecular and Cellular Probes. *Dalton Trans.* **2012**, *41*, 6021. [CrossRef]
4. Min, J.; Zhang, Q.; Sun, W.; Cheng, Y.; Wang, L. Neutral Copper(i) Phosphorescent Complexes from Their Ionic Counterparts with 2-(2′-Quinolyl)Benzimidazole and Phosphine Mixed Ligands. *Dalton Trans.* **2011**, *40*, 686–693. [CrossRef]
5. Zhang, Q.; Ding, J.; Cheng, Y.; Wang, L.; Xie, Z.; Jing, X.; Wang, F. Novel Heteroleptic Cu1 Complexes with Tunable Emission Color for Efficient Phosphorescent Light-Emitting Diodes. *Adv. Funct. Mater.* **2007**, *17*, 2983–2990. [CrossRef]

6. Takeda, H.; Kobayashi, A.; Tsuge, K. Recent Developments of Photoactive Cu(I) and Ag(I) Complexes with Diphosphine and Related Ligands. *Coord. Chem. Rev.* **2022**, *470*, 214700. [CrossRef]
7. Shafikov, M.Z.; Czerwieniec, R.; Yersin, H. Ag(i) Complex Design Affording Intense Phosphorescence with a Landmark Lifetime of over 100 Milliseconds. *Dalton Trans.* **2019**, *48*, 2802–2806. [CrossRef]
8. Shafikov, M.Z.; Suleymanova, A.F.; Czerwieniec, R.; Yersin, H. Thermally Activated Delayed Fluorescence from Ag(I) Complexes: A Route to 100% Quantum Yield at Unprecedentedly Short Decay Time. *Inorg. Chem.* **2017**, *56*, 13274–13285. [CrossRef]
9. Zuideveld, M.A.; Swennenhuis, B.H.G.; Boele, M.D.K.; Guari, Y.; Van Strijdonck, G.P.F.; Reek, J.N.H.; Kamer, P.C.J.; Goubitz, K.; Fraanje, J.; Lutz, M.; et al. The Coordination Behaviour of Large Natural Bite Angle Diphosphine Ligands towards Methyl and 4-Cyanophenylpalladium(II) Complexes. *J. Chem. Soc. Dalton Trans.* **2002**, *11*, 2308–2317. [CrossRef]
10. Freixa, Z.; Van Leeuwen, P.W.N.M. Bite Angle Effects in Diphosphine Metal Catalysts: Steric or Electronic? *Dalton Trans.* **2003**, *10*, 1890–1901. [CrossRef]
11. Flecken, F.; Grell, T.; Hanf, S. Transition Metal Complexes of the PPO/POP Ligand: Variable Coordination Chemistry and Photo-Luminescence Properties. *Dalton Trans.* **2022**, *51*, 8975–8985. [CrossRef]
12. Zhao, Y.H.; Li, H.Y.; Young, D.J.; Cao, X.; Zhu, D.L.; Ren, Z.G.; Li, H.X. Heteroleptic Copper(i) Complexes [Cu(Dmp)(N^P)]BF$_4$ for Photoinduced Atom-Transfer Radical Addition Reactions. *Dalton Trans.* **2023**, *52*, 8142–8154. [CrossRef]
13. Beaudelot, J.; Oger, S.; Peruško, S.; Phan, T.A.; Teunens, T.; Moucheron, C.; Evano, G. Photoactive Copper Complexes: Properties and Applications. *Chem. Rev.* **2022**, *122*, 16365–16609. [CrossRef]
14. Glykos, D.; Plakatouras, J.C.; Malandrinos, G. Bis(2-Phenylpyridinato,-C2′,N)[4,4′-Bis(4-Fluorophenyl)-6,6′-Dimethyl-2,2′-Bipyridine] Iridium(III) Hexafluorophosphate. *Molbank* **2023**, *2023*, 4–11. [CrossRef]
15. Kouvatsis, P.; Glykos, D.; Plakatouras, J.C.; Malandrinos, G. [6-(Furan-2-Yl)-2,2′-Bipyridine]Bis(Triphenylphosphine) Copper(I) Tetrafluoroborate. *Molbank* **2023**, *2023*, M1724. [CrossRef]
16. Glykos, D.; Plakatouras, J.C.; Malandrinos, G. [4,4′-Bis(4-Fluorophenyl)-6,6′-Dimethyl-2,2′-Bipyridine] [Bis (2-(Diphenylphosphino) Phenyl) Ether] Silver(I) Hexafluorophosphate. *Molbank* **2023**, *2023*, M1675. [CrossRef]
17. Glykos, D.; Plakatouras, J.C.; Malandrinos, G. Solution-State Studies, X-Ray Structure Determination and Luminescence Properties of an Ag(I) Heteroleptic Complex Containing 2,9-Bis(Styryl)-1,10-Phenanthroline Derivative and Triphenylphosphine. *Inorganics* **2023**, *11*, 467. [CrossRef]
18. Kouvatsis, P.; Glykos, D.; Plakatouras, J.C.; Malandrinos, G. [6-(Thiophen-2-Yl)-2,2′-Bipyridine]Bis(Triphenylphosphine) Copper(I) Tetrafluoroborate. *Molbank* **2023**, *2023*, M1605. [CrossRef]
19. Bonaccorso, C.; Cesaretti, A.; Elisei, F.; Mencaroni, L.; Spalletti, A.; Fortuna, C.G. New Styryl Phenanthroline Derivatives as Model D−π−A−π−D Materials for Non-Linear Optics. *ChemPhysChem* **2018**, *19*, 1917–1929. [CrossRef]
20. Sun, L.Z.; Kuang, X.N.; Lin, S.; Zhao, L.; Yu, X.; Li, Z.F.; Liu, M.; Xin, X.L.; Yang, Y.P.; Jin, Q.H. Nine Heteroleptic Copper(I)/Silver(I) Complexes Prepared from Phosphine and Diimine Ligands: Syntheses, Structures and Terahertz Spectra. *Polyhedron* **2020**, *175*, 114177. [CrossRef]
21. Zhang, Y.R.; Cui, Y.Z.; Jin, Q.H.; Yang, Y.P.; Liu, M.; Li, Z.F.; Bi, K.L.; Zhang, C.L. Syntheses, Structural Characterizations and Terahertz Spectra of Ag(I)/Cu(I) Complexes with Bis[2-(Diphenylphosphino)Phenyl]Ether and N^N Ligands. *Polyhedron* **2017**, *122*, 86–98. [CrossRef]
22. Wu, Z.; Cui, S.; Zhao, Z.; He, B.; Li, X.L. Photophysical Properties of Homobimetallic Cu(i)-Cu(i) and Heterobimetallic Cu(i)-Ag(i) Complexes of 2-(6-Bromo-2-Pyridyl)-1H-Imidazo[4,5-f][1,10]Phenanthroline. *New J. Chem.* **2022**, *46*, 8881–8891. [CrossRef]
23. Kaeser, A.; Delavaux-Nicot, B.; Duhayon, C.; Coppel, Y.; Nierengarten, J.F. Heteroleptic Silver(I) Complexes Prepared from Phenanthroline and Bis-Phosphine Ligands. *Inorg. Chem.* **2013**, *52*, 14343–14354. [CrossRef]
24. Moudam, O.; Tsipis, A.C.; Kommanaboyina, S.; Horton, P.N.; Coles, S.J. First Light-Emitting Electrochemical Cell with [Ag(i)(N^N)(P^P)] Type Complex. *RSC Adv.* **2015**, *5*, 95047–95053. [CrossRef]
25. Wang, Y.; Kuang, X.N.; Cui, Y.Z.; Xin, X.L.; Han, H.L.; Liu, M.; Yang, Y.P.; Jin, Q.H. Synthesis, Structure, Luminescent Properties, and Photocatalytic Behavior of 0D–3D Silver(I) Complexes Bearing Both Diphosphine Ligands and 1,10-Phenanthroline Derivatives. *Polyhedron* **2018**, *155*, 135–143. [CrossRef]
26. Lipinski, S.; Cavinato, L.M.; Pickl, T.; Biffi, G.; Pöthig, A.; Coto, P.B.; Fernández-Cestau, J.; Costa, R.D. Dual-Phosphorescent Heteroleptic Silver(I) Complex in Long-Lasting Red Light-Emitting Electrochemical Cells. *Adv. Opt. Mater.* **2023**, *11*, 2203145. [CrossRef]
27. Teng, T.; Li, K.; Cheng, G.; Wang, Y.; Wang, J.; Li, J.; Zhou, C.; Liu, H.; Zou, T.; Xiong, J.; et al. Lighting Silver(I) Complexes for Solution-Processed Organic Light-Emitting Diodes and Biological Applications via Thermally Activated Delayed Fluorescence. *Inorg. Chem.* **2020**, *59*, 12122–12131. [CrossRef]
28. Nemati Bideh, B.; Shahroosvand, H.; Nazeeruddin, M.K. High-Efficiency Deep-Red Light-Emitting Electrochemical Cell Based on a Trinuclear Ruthenium(II)–Silver(I) Complex. *Inorg. Chem.* **2021**, *60*, 11915–11922. [CrossRef]
29. Beliaeva, M.; Belyaev, A.; Grachova, E.V.; Steffen, A.; Koshevoy, I.O. Ditopic Phosphide Oxide Group: A Rigidifying Lewis Base to Switch Luminescence and Reactivity of a Disilver Complex. *J. Am. Chem. Soc.* **2021**, *143*, 15045–15055. [CrossRef]
30. Wang, C.; Li, Z. Molecular Conformation and Packing: Their Critical Roles in the Emission Performance of Mechanochromic Fluorescence Materials. *Mater. Chem. Front.* **2017**, *1*, 2174–2194. [CrossRef]
31. Yuan, M.-S.; Du, X.; Xu, F.; Wang, D.-E.; Wang, W.-J.; Li, T.-B.; Tu, Q.; Zhang, Y.; Du, Z.; Wang, J. Aggregation-Induced Bathochromic Fluorescent Enhancement for Fluorenone Dyes. *Dye Pigment* **2015**, *123*, 355–362. [CrossRef]

32. Jia, J.H.; Liang, D.; Yu, R.; Chen, X.L.; Meng, L.; Chang, J.F.; Liao, J.Z.; Yang, M.; Li, X.N.; Lu, C.Z. Coordination-Induced Thermally Activated Delayed Fluorescence: From Non-TADF Donor-Acceptor-Type Ligand to TADF-Active Ag-Based Complexes. *Chem. Mater.* **2020**, *32*, 620–629. [CrossRef]
33. Balakrishna, M.S.; Venkateswaran, R.; Mobin, S.M. Mixed-Ligand Silver(I) Complexes Containing Bis[2-(Diphenylphosphino)Phenyl]Ether and Pyridyl Ligands. *Inorganica Chim. Acta* **2009**, *362*, 271–276. [CrossRef]
34. Cui, Y.Z.; Yuan, Y.; Han, H.L.; Li, Z.F.; Liu, M.; Jin, Q.H.; Yang, Y.P.; Zhang, Z.W. Synthesis, Characterization, and Luminescent Properties of Silver(I) Complexes Based on Diphosphine Ligands and 2,9-Dimethyl-1,10-Phenanthroline. *Z. Anorg. Allg. Chem.* **2016**, *642*, 953–959. [CrossRef]
35. Wang, Y.; Cui, Y.Z.; Li, Z.F.; Liu, M.; Yang, Y.P.; Zhang, Z.W.; Xin, X.L.; Jin, Q.H. Synthesis, Characterization, and Luminescent Properties of Silver(I) Complexes Based on Diphosphine Ligands and 6,7-Dicyanodipyridoquinoxaline. *Z. Anorg. Allg. Chemie* **2017**, *643*, 1253–1261. [CrossRef]
36. Ishida, H.; Tobita, S.; Hasegawa, Y.; Katoh, R.; Nozaki, K. Recent Advances in Instrumentation for Absolute Emission Quantum Yield Measurements. *Coord. Chem. Rev.* **2010**, *254*, 2449–2458. [CrossRef]
37. *APEX 3*; SAINT, SHELXT 2016; Bruker AXS Inc.: Madison, WI, USA, 2016.
38. Sheldrick, G.M. *SADABS 1996*; University of Göttingen: Göttingen, Germany.
39. Sheldrick, G.M. Crystal Structure Refinement with SHELXL. *Acta Crystallogr. Sect. C Struct. Chem.* **2015**, *71*, 3–8. [CrossRef]
40. Hübschle, C.B.; Sheldrick, G.M.; Dittrich, B. ShelXle: A Qt Graphical User Interface for SHELXL. *J. Appl. Crystallogr.* **2011**, *44*, 1281–1284. [CrossRef]
41. Spek, A.L. Structure Validation in Chemical Crystallography. *Acta Crystallogr. Sect. D Biol. Crystallogr.* **2009**, *65*, 148–155. [CrossRef]
42. Barbour, L.J. X-Seed—A Software Tool for Supramolecular Crystallography. *J. Supramol. Chem.* **2001**, *1*, 189–191. [CrossRef]
43. Frisch, M.J.; Trucks, G.W.; Schlegel, H.B.; Scuseria, G.E.; Robb, M.A.; Cheeseman, J.R.; Scalmani, G.; Barone, V.; Petersson, G.A.; Nakatsuji, H.; et al. *Gaussian 16W*, Revision C.01; Gaussian, Inc.: Wallingford, CT, USA, 2016.
44. Vetere, V.; Adamo, C.; Maldivi, P. Performance of the 'parameter free' PBE0 functional for the modeling of molecular properties of heavy metals. *Chem. Phys. Lett.* **2000**, *325*, 99–105. [CrossRef]
45. Adamo, C.; Barone, V. Inexpensive and accurate predictions of optical excitations in transition-metal complexes: The TDDFT/PBE0 route. *Theor. Chem. Acc.* **2000**, *105*, 169–172. [CrossRef]
46. Adamo, C.; Barone, V. Toward reliable density functional methods without adjustable parameters: The PBE0 model. *J. Chem. Phys.* **1999**, *110*, 6158–6170. [CrossRef]
47. Ernzerhof, M.; Scuseria, G.E. Assessment of the Perdew–Burke–Ernzerhof exchange-correlation functional. *J. Chem. Phys.* **1999**, *110*, 5029–5036. [CrossRef]
48. Adamo, C.; Scuseria, G.E.; Barone, V. Accurate excitation energies from time-dependent density functional theory: Assessing the PBE0 model. *J. Chem. Phys.* **1999**, *111*, 2889–2899. [CrossRef]
49. Adamo, C.; Barone, V. Toward reliable adiabatic connection models free from adjustable parameters. *Chem. Phys. Lett.* **1997**, *274*, 242–250. [CrossRef]
50. Tomasi, J.; Mennucci, B.; Cammi, R. Quantum Mechanical Continuum Solvation Models. *Chem. Rev.* **2005**, *105*, 2999–3093. [CrossRef]

Disclaimer/Publisher's Note: The statements, opinions and data contained in all publications are solely those of the individual author(s) and contributor(s) and not of MDPI and/or the editor(s). MDPI and/or the editor(s) disclaim responsibility for any injury to people or property resulting from any ideas, methods, instructions or products referred to in the content.

Article

Different Patterns of Pd-Promoted C-H Bond Activation in (Z)-4-Hetarylidene-5(4H)-oxazolones and Consequences in Photophysical Properties

Miguel Martínez, David Dalmau, Olga Crespo, Pilar García-Orduña, Fernando Lahoz, Antonio Martín and Esteban P. Urriolabeitia *

Instituto de Síntesis Química y Catálisis Homogénea, CSIC-Universidad de Zaragoza, 50009 Zaragoza, Spain; 780460@unizar.es (M.M.); ddalmau@unizar.es (D.D.); ocrespo@unizar.es (O.C.); mpgaror@unizar.es (P.G.-O.); lahoz@unizar.es (F.L.); tello@unizar.es (A.M.)
* Correspondence: esteban.u.a@csic.es or esteban@unizar.es

Citation: Martínez, M.; Dalmau, D.; Crespo, O.; García-Orduña, P.; Lahoz, F.; Martín, A.; Urriolabeitia, E.P. Different Patterns of Pd-Promoted C-H Bond Activation in (Z)-4-Hetarylidene-5(4H)-oxazolones and Consequences in Photophysical Properties. *Inorganics* **2024**, *12*, 271. https://doi.org/10.3390/inorganics12100271

Academic Editor: Binbin Chen

Received: 17 September 2024
Revised: 15 October 2024
Accepted: 17 October 2024
Published: 18 October 2024

Copyright: © 2024 by the authors. Licensee MDPI, Basel, Switzerland. This article is an open access article distributed under the terms and conditions of the Creative Commons Attribution (CC BY) license (https://creativecommons.org/licenses/by/4.0/).

Abstract: This work aims to amplify the fluorescence of (Z)-4-hetarylidene-5(4H)-oxazolones **1** by suppression of the hula-twist non-radiative deactivation pathway by C^N-orthopalladation of the 4-hetarylidene ring. Different (Z)-4-hetarylidene-2-phenyl-5(4H)-oxazolones, **1a–1c**, prepared by the Erlenmeyer–Plöchl method, have been studied. The orthopalladation of (Z)-2-phenyl-4-(5-thiazolylmethylene)-5(4H)-oxazolone (**1a**) takes place by C-H bond activation of the H4 of the heterocycle and C^N-chelation, giving the dinuclear trifluoroacetate derivative **2a**. By further metathesis of bridging ligands in **2a**, complexes containing the orthometalated oxazolone and a variety of ligands **3a–5a,** were prepared. The study of the photophysical properties of **1a–5a** shows that the bonding of the Pd metal to the 4-hetaryliden-5(4H)-oxazolone does not promote, in these cases, an increase in fluorescence. Interestingly, the orthopalladation of (Z)-2-phenyl-4-(4-thiazolylmethylene)-5(4H)-oxazolone (**1b**) gives orthopalladated **2b**, where the incorporation of the Pd to the oxazolone takes place by C-H bond activation of the ortho-H2 of the 2-phenyl group, ring opening of the oxazolone heterocycle and simultaneous N,N-bonding of the N atoms of the thiazole ring and the generated benzamide fragment. This N^N^C-tridentate dianionic bonding mode is obtained for the first time in oxazolones. Despite a similar lock of the hula-twist deactivation, **2b** does not show fluorescence.

Keywords: oxazolones; fluorescence; heterocycles; C-H bond activation; palladium

1. Introduction

The synthesis of compounds with photophysical properties of interest, such as fluorescence or phosphorescence, is a well-established area of research. However, in recent years, it has been experiencing growing development due to its multiple applications. The most impactful and well-known application is their use as screens in optical devices, followed by their employment as biological markers or probes for disease detection through the visualization of the vital activity of cells and organisms [1–7].

There are various criteria to consider when designing the synthesis of a molecule with potential luminescent properties. In this regard, nature provides us with models to follow, as many natural compounds exhibit luminescent properties. A notable example is the GFP (Green Fluorescent Protein) and its derivatives. GFP has come to be commonly known as the microscope of the 21st century, due to its ease of visualizing cellular internal functioning when irradiated with UV light, which causes the emission of an intense green light. The utility of GFP and its derivatives was so significant that, in 2008, the Nobel Prize in Chemistry was awarded to Professors Shimomura, Chalfie, and Tsien [8–10].

However, the true reason for the photophysical activity of GFP is the protein's chromophore, a small molecule whose backbone contains the (Z)-4-arylidene-5(4H)-imidazolone unit (Figure 1a), confined in the internal part of the protein in a very rigid structural environment. These GFP chromophores are formed by the internal modification of some amino acid residues. These modifications are catalyzed by the fluorescent protein itself and do not require external enzymatic activity. Recently, it has been demonstrated that GFP exhibits a wide spectral variety due to the presence of different chromophore structures [11–14].

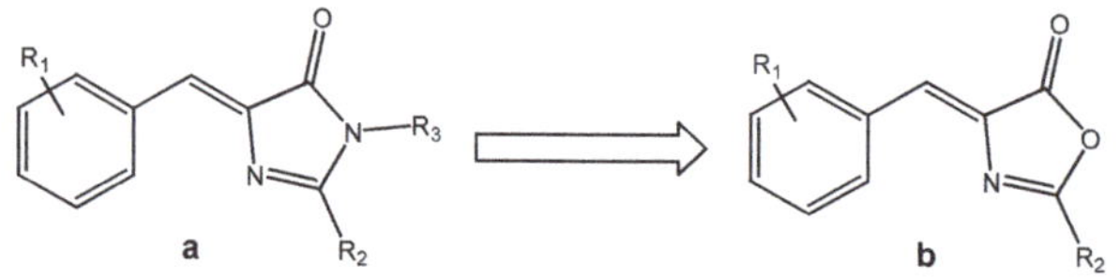

Figure 1. Structure of: (**a**) 4-aryliden-5(4*H*)-imidazolone; (**b**) 4-aryliden-5(4*H*)-oxazolone.

The 4-arylidene-5(4*H*)-oxazolone compounds, shown in Figure 1b, are structural analogs of the 4-arylidene-5(4*H*)-imidazolones and can be considered their synthetic precursors, as the imidazolones can be easily obtained from the oxazolones. Both types of compounds, oxazolones and imidazolones, are very versatile precursors in the synthesis of α-amino acids and, as expected, they also exhibit interesting photophysical properties [15]. However, it is important to note that the existence of photophysical properties of interest in the chromophores of the GFP protein is a direct consequence of the rigid environment they have inside the protein. In the absence of this rigidity, for example, in a solution outside of the protein, the luminescence disappears due to alternative relaxation processes through non-radiative pathways, known as "hula-twist" (Figure 2a) [16–21].

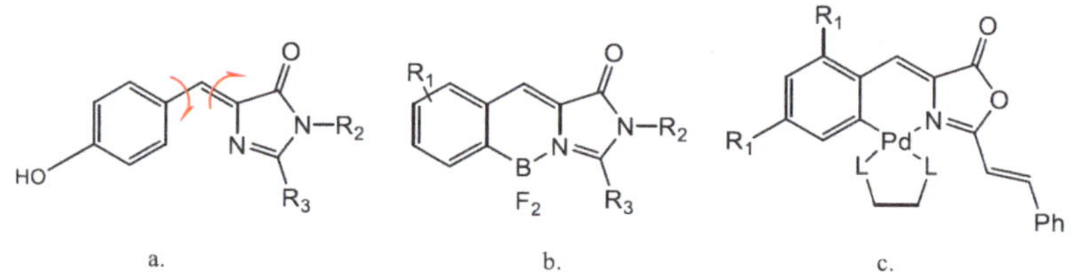

Figure 2. (**a**) "Hula-twist" relaxation (Imidazolone); (**b**) Chromophore with conformational blocking through the introduction of the BF$_2$ fragment (Imidazolone); (**c**) Chromophore with conformational blocking through PdL$_2$ (Oxazolone).

If these structures exhibit high fluorescence in a very rigid and complex environment, such as that of the protein, it is important to prevent the loss of fluorescence in non-rigid environments. Several solutions have been proposed to prevent the loss of fluorescence through non-radiative pathways, all centered on imidazolones. The most interesting of these solutions is the conformational blocking of the arylidene fragment of the heterocycle by anchoring the ortho carbon atom and the nitrogen of the heterocycle through coordination to the same molecular fragment [22,23]. The insertion of BF$_2$, as shown in Figure 2b, has provided spectacular results in increasing fluorescence, as it decreases the degrees of freedom and increases conformational restriction, thus minimizing the effects of non-radiative losses through hula-twist [22,23].

Subsequently, it has been shown that it is also possible to achieve this objective using palladium as an anchor, both in oxazolones and imidazolones, as shown in Figure 2c [15,24–26]. The incorporation of Pd into the oxazolone and imidazolone skeletons occurs easily through C-H bond activation. The square-planar environment of palladium additionally requires the presence of auxiliary ligands, whose steric and electronic properties also influence fluorescence. Using this type of strategy with Pd organometallics, fluorescence amplifications of up to two orders of magnitude have been obtained in certain cases when transitioning from the free ligand to the complex, with quantum yields of up to 28% [25]. All the photophysical results discussed so far have been carried out on orthopalladated oxazolones containing a substituted 4-arylidene group. However, no study has been conducted on oxazolones that contain heterocycles in that position, that is, 4-hetarylidene groups as represented in Figure 3 (X, Y, Z in Figure 3 can be S, O, NH, CH, etc.). In particular, heterocycles such as thiazole or imidazole could be of special interest due to their well-known intrinsic luminescent properties.

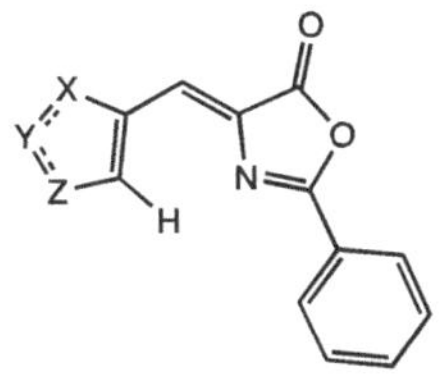

Figure 3. General structure of (Z)-4-hetaryliden-5(4H)-oxazolones (X, Y, and Z can be O, NH, or S).

Given the critical role of the arylidene fragment in the photophysical properties of oxazolones, this work addresses the synthesis and characterization of (Z)-4-hetarylidene-2-phenyl-5(4H)-oxazolones, the study of their reactivity towards Pd(II), and especially the measurement of the photophysical properties of both the free ligands and the complexes, aiming to determine if fluorescence amplification occurs after incorporating Pd into the oxazolone skeleton. It is particularly noteworthy that the study of the reactivity of particular (Z)-4-hetarylidene-2-phenyl-5(4H)-oxazolones has led to the characterization of a new type of C-H bond activation in these substrates.

2. Results and Discussion

2.1. Synthesis and Characterization of Oxazolones from Heterocycles

The oxazolones **1a–1c**, shown in Figure 4, were prepared from the corresponding aldehydes following the Erlenmeyer–Plöchl method described in the literature [27–33]. In the case of oxazolone **1c**, additional N-acetylation of the NH bond of the imidazole ring has been observed due to its reaction with the acetic anhydride used as reaction solvent. The presence of this acetyl group should not limit further reactivity with palladium and confers additional stability to this ring. All oxazolones **1** were obtained as yellow solids stable to air and moisture, with yields ranging from 39% to 97%. They have been selected aiming to cover various structural situations, such as N,S-heterocycles versus N,N-heterocycles and within N,S-heterocycles; we change the reactive position to observe the influence of the substitution position on the ring.

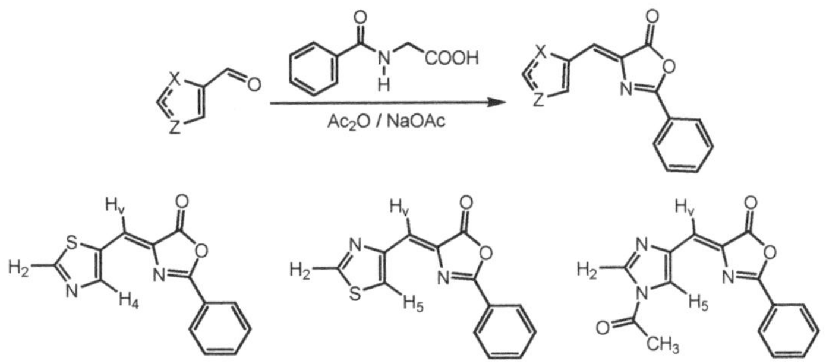

Figure 4. General synthesis of oxazolones and examples **1a–1c** employed in this work (yields in parentheses).

The characterization of **1a–1c** has been carried out by HRMS and NMR methods (see Materials and Methods section and Supplementary Materials). The HRMS spectra in all cases show the presence of a peak with the isotopic distribution corresponding to the molecular ion [M]$^+$, which has captured Na$^+$ [M + Na]$^+$ or K$^+$ [M + K]$^+$ cations. These results confirm the stoichiometries shown in Figure 4. The ^{1}H and ^{13}C NMR spectra of **1a–1c** show characteristic signals that suggest the structures of the oxazolones depicted in Figure 4. The ^{1}H-NMR spectrum of **1a** shows two doublets of doublets at 9.05 and 8.32 ppm, assigned to the hydrogens of the 5-thiazole heterocycle, H2 and H4, respectively. Meanwhile, the H2 and H5 protons of the 4-thiazolyl group in **1b** appear as two doublets at 8.88 and 8.80 ppm, and those corresponding to the imidazole group in **1c** appear at 8.54 ppm (H2) and 8.21 ppm (H5). The ^{1}H-NMR spectrum of **1c** also shows the presence of a singlet at 2.70 ppm, integrating for 3H. This singlet correlates in the ^{1}H-^{13}C HMBC with a carbonyl group at 165.88 ppm and with a methyl group at 22.70 ppm in the ^{1}H-^{13}C HSQC. These findings confirm the presence of the acetyl group –C(O)Me in the molecule. Signals due to H belonging to the Ph group and the vinyl proton appear as expected. Similarly, the ^{13}C{^{1}H} NMR spectrum of **1a** shows peaks at 159.62, 149.26, and 132.28 ppm, which are characteristic of the C2, C4, and C5 carbons, respectively, of the 5-substituted thiazole ring. The formation of the oxazolone heterocycle is inferred from the observation of signals at 165.82 ppm, assigned to the C=O group, 163.47 ppm (C=N group), and a third at 133.22 ppm due to the quaternary carbon of the arylidene group. The signal corresponding to the vinylic carbon is located at 120.97 ppm. This analysis is similarly applicable to **1b** and **1c**.

The determination of the molecular structure of oxazolone **1c** by X-ray diffraction, depicted in Figure 5, provides additional structural information. Crystals of compound **1c** were obtained by slow diffusion of *n*-pentane into a solution of the crude compound in CH$_2$Cl$_2$ at −18 °C over the course of one week. The Supplementary Materials includes the main crystallographic data related to data collection, and the process of solving and refining **1c**.

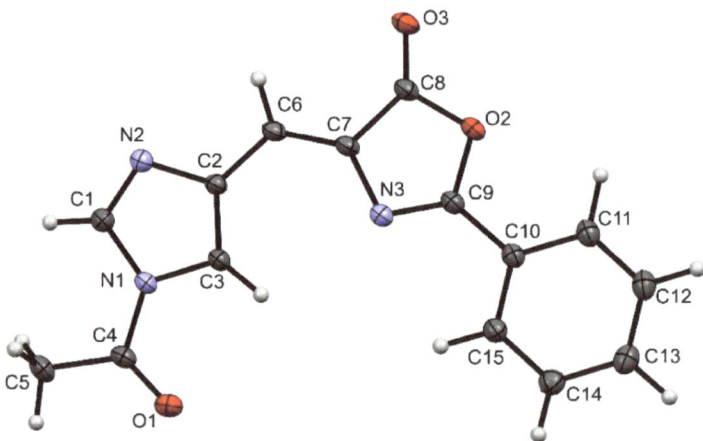

Figure 5. Molecular structure of oxazolone **1c**. Thermal ellipsoids are drawn at 50% probability level.

The structure clearly shows the formation of the 5(4*H*)-oxazolone ring, substituted at position 2 by a phenyl group and at position 4 by an *N*-acetyl-imidazolidene group, which in turn confirms the substitution of the NH proton of the imidazole group by an acetyl group, as the NMR spectra suggested. The configuration of the C=C double bond is (*Z*), as is typical in oxazolones [34,35]. A comparison of bond lengths and angles of this oxazolone with values found for other unsaturated oxazolones reported in the literature shows that the presence of the heterocycle on the exocyclic vinylic carbon does not cause deviations in the oxazolone ring [36].

2.2. C-H Bond Activation Reactions: Synthesis and Characterization of the Orthometalated Dinuclear Pd Complex with Trifluoroacetate Bridges (**2a**) and the Tridentate Complex (**2b**)

Once the oxazolones were characterized, their reactivity with Pd(OAc)$_2$ was tested with the aim of incorporating Pd into the molecular skeleton of the oxazolone and restricting possible fluorescence deactivation pathways. The incorporation of Pd into the oxazolone skeleton **1a** is achieved by orthometalation through a C-H activation process, described in the literature [24–26,36–39], which generates the dinuclear complex with trifluoroacetate bridges **2a** (55% yield) shown in Scheme 1.

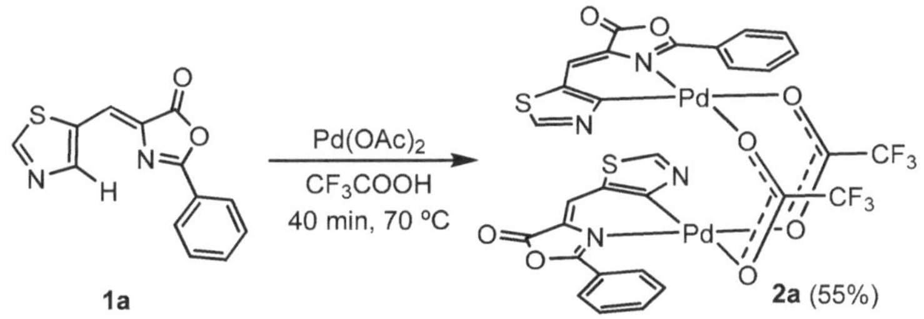

Scheme 1. Synthesis of dinuclear orthopalladated complex **2a** through CH bond activation.

Heating the mixture of **1a** and Pd(OAc)$_2$ in trifluoroacetic acid at 70 °C causes the initial suspension to change color to a more intense brown. After 40 min heating, the addition of water to the cool suspension causes more solid to precipitate. This solid is filtered and washed with water to remove the excess of acid. This solid contains complex **2a** contaminated with black Pd^0. Therefore, it is then recrystallized by adding CH$_2$Cl$_2$ and filtering the resulting suspension through Celite. The resulting clear yellow solution is evaporated to dryness, yielding **2a** as a yellow solid with a 55% yield.

The IR spectrum of **2a** shows the C-O and the C-F stretching vibrations at 1627 cm^{-1} and at 1139 cm^{-1}, respectively, showing the presence of the anionic CF$_3$CO$_2$ ligand. The ^{1}H NMR spectrum shows the disappearance of the signal assigned to H4 in the 5-thiazole fragment, while the signals assigned to the vinyl hydrogen and the phenyl ring remain unchanged, indicating that metallation has selectively occurred on the thiazole ring. These data suggest that the oxazolone coordinates as a C^N-chelating ligand and that the coordination sphere of Pd is completed with the trifluoroacetate group. Due to this, and considering the behavior observed in other structurally related oxazolones, the formation of **2a** is proposed as a dinuclear complex with the carboxylate ligands acting as O,O'-bridges. Furthermore, the relative arrangement of the two oxazolone chelate fragments must be *transoid*, as inferred from the observation of a single singlet in the ^{19}F NMR spectrum of **2a**, corresponding to the two chemically equivalent CF$_3$ fragments, as shown in Scheme 1.

Similarly, the reactivity of oxazolones **1b** and **1c** with Pd(OAc)$_2$ in CF$_3$CO$_2$H was tested under a variety of conditions. All attempts to obtain an orthometalated product from oxazolone **1c** were unsuccessful, with its decomposition being observed in both trifluoroacetic acid and acetic acid, even at room temperature. The reactivity of **1b** with Pd(OAc)$_2$ was more successful, although the result was surprising. Thus, treatment of **1b** with Pd(OAc)$_2$ (1:1 molar ratio) in CF$_3$CO$_2$H at 70 °C for 2 h led to the formation of a deep red solid, completely insoluble in the usual solvents used for NMR, including DMSO. However, this red solid, when suspended in CD$_2$Cl$_2$, reacts with excess pyridine-d$_5$, causing a color change from red to pale yellow and the almost immediate dissolution of the initial suspension. This behavior is typical of Pd systems that maintain carboxylate or halide bridges and is used to characterize highly insoluble orthometalated systems by transforming them into more soluble ones [15,24–26].

The spectroscopic characterization (^{1}H NMR) of this solution shows a mixture of at least two products (see Supplementary Materials) with the same signal pattern but different chemical shifts. The ^{1}H NMR spectrum shows peaks that suggest the presence of a C_6H_4 spin system with two ortho substituents, as well as the presence of the two protons from the thiazole group. Both findings indicate that in oxazolone **1b**, the C-H activation has occurred on the C_6H_5 ring and not on the heterocycle as in the case of **1a**. Incidentally, the activation of C-H bonds in the 4-arylidene ring is much more common than in the 2-aryl ring; therefore, this compound represents a quite rare example of alternative C-H activation in oxazolones [36–39]. The moderate solubility of this compound, even in CD_2Cl_2, prevented the acquisition of high-quality ^{13}C spectra, making it impossible to advance the structural characterization beyond this observation. However, the slow evaporation of these solutions in CD_2Cl_2 led to the formation of a small number of yellow crystals. The resolution of the crystal structure by X-ray diffraction of these crystals did provide relevant information about the course of the reaction.

The structural determination shows that the crystals result from the co-crystallization of an orthopalladated unit (shown in Figure 6) together with the complex $[Pd(py)_4](CF_3CO_2)_2$, and that they also contain CH_2Cl_2. This mixture of palladium complexes crystallizes in the triclinic system, space group P-1, and the asymmetric unit of each unit cell contains one orthometalated complex unit, half of a $[Pd(py)_4]^{2+}$ complex, one CF_3CO_2 anion, and 0.4 molecules of dichloromethane. Various ORTEP diagrams of the contents of the asymmetric unit are shown in the Supplementary Materials, while Figure 6 shows an ORTEP diagram of the complex derived from the orthopalladation of oxazolone **1b**.

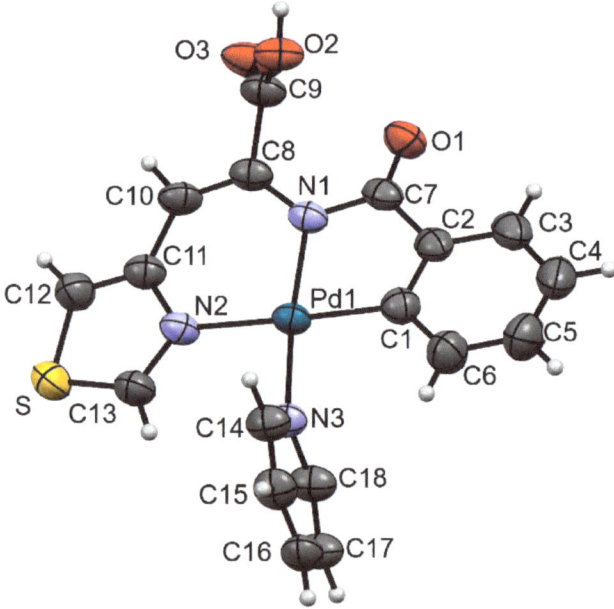

Figure 6. Molecular structure of palladacycle **2b**. Thermal ellipsoids are drawn at 50% probability level.

Figure 6 confirms that, indeed, the C-H activation has occurred at the ortho position of the 2-phenyl ring, while the thiazole group has not undergone any activation of bonds and is simply N-coordinated to the Pd atom. It also shows that the oxazolone ring has undergone a significant modification, as it has been transformed into a N-benzoyl-dehydroamino acid by the opening of the oxazolone ring through hydrolysis and the generation of the carboxylic acid groups C8-C9(O3)O2-H and benzamide N1-C7(O1)-(C1-C6). In addition to the σ(Pd1-C1) bond that formally is anionic and saturates charge, deprotonation of the benzamide-type N1 has occurred, generating an amide anion that also saturates charge. The coordination of N2 of the thiazole ring causes the dehydroamino acid ligand generated from the oxazolone to act as a κ^3-N,N,C-tridentate ligand. The coordination sphere of Pd is completed by the coordination of a pyridine ligand through nitrogen N3. Beyond the significant transformation that the original oxazolone **1b** has undergone to become the tridentate ligand shown in Figure 6, the analysis of bond distances and angles in this complex and in the co-crystallized species [Pd(py)$_4$](CF$_3$CO$_2$)$_2$ does not show significant deviations from the values found in the literature for related species [40].

According to these results, we propose that the formation of **2b** takes place following the process shown in Scheme 2.

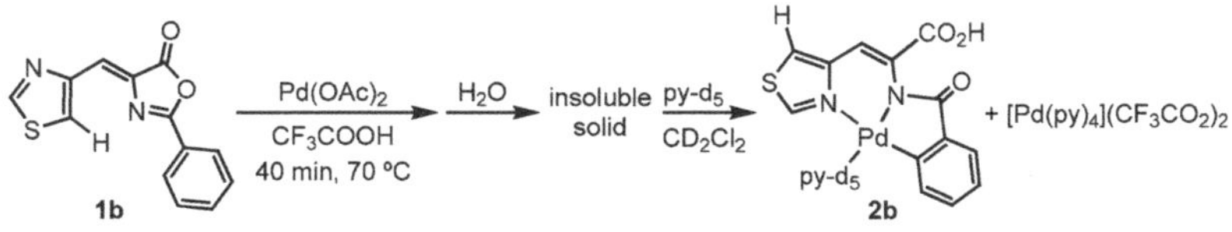

Scheme 2. Alternative C-H activation of oxazolone **1b** and synthesis of **2b**.

Clearly, it is necessary to consider that this characterization corresponds to only a portion of the original sample, as this is only one of the ortho-metalated complexes obtained from the crude compound. Currently, we are focusing on the characterization of the insoluble material.

Regarding the process that leads to the formation of the ortho-metalated complex, the source of the water involved in the hydrolysis reaction and ring-opening could be that contained in the trifluoroacetic acid (the reaction solvent), so the insoluble solid might already contain the hydrolyzed oxazolone ring, or it could be that it was added later for washing. On the other hand, it has not been possible to determine unequivocally whether the C-H activation and ortho-palladation occur after the opening of the oxazolone, that is, in the dehydroamino acid species, or before the opening, in which case it would be an attempt by the system to achieve additional stabilization. Additional control experiments show that it is more plausible to be the latter case, as attempts to orthopalladate the corresponding dehydroamino acid lead to the recovery of the starting products. Similarly, we have observed that in 2 + 2 photodimerization reactions of oxazolones to form cyclobutanes, the opening of the heterocycle and the formation of the corresponding amide esters is a common occurrence, which significantly stabilizes the final cyclobutanes by removing the strain in the oxazolone ring [41]. Therefore, we propose initial C-H activation on the oxazolone, formation of the orthopalladated tridentate complex, and finally, hydrolysis of the oxazolone ring to give the metalated species shown in Figure 6 and Scheme 2.

The differing reactivity of **1a** and **1b** towards Pd(OAc)$_2$, despite their structural similarities, is striking. Our proposal is that the different coordinating abilities of the heteroatoms in the thiazole ring (S vs. N) with respect to palladium could be responsible for this switch in reactivity towards the 2-phenyl ring. In the case of oxazolone **1a**, coordination of the sulfur atom of the thiazole could occur as an alternative to C-H activation. However, sulfur coordination to Pd is inherently weak, and under reaction conditions (70 °C), the Pd-S bond would be very unstable. This could lead to rotation of the thiazole and eventual activation of the C-H bond at position 4 of the thiazole, generating **2a**. In contrast, in oxazolone **1b**, coordination of the nitrogen can occur and may compete with the activation of the bond at position 5 of the thiazole. Furthermore, after coordination of both the nitrogen of the thiazole and the nitrogen of the oxazolone, a stable chelate is formed, which strongly directs C-H activation to the ortho position of the 2-phenyl ring and ultimately leads to the observed ortho-palladation. We do not have a clear explanation for the lack of reactivity of oxazolone **1c**, beyond the fact that the presence of the strongly electron-withdrawing MeC(O) group on the nitrogen of the imidazole may deactivate this ring.

2.3. Reactivity of (2a), Synthesis of 3a–5a

In order to obtain a greater variety of Pd complexes with ortho-metalated oxazolone **1a** for measuring their photophysical properties, the reactivity of **2a** has been tested in ligand metathesis and bridge cleavage reactions. Dimers with chloride bridges are highly reactive and suitable for preparing a wide range of mononuclear complexes by adding ligands that can act as chelates or nucleophiles to break the chloride bridges. The synthesis of dimer **3a** is carried out from **2a** and the excess of LiCl through a metathesis process where trifluoroacetate bridges are replaced by chloride bridges, as described in the literature and represented in Scheme 3 [24].

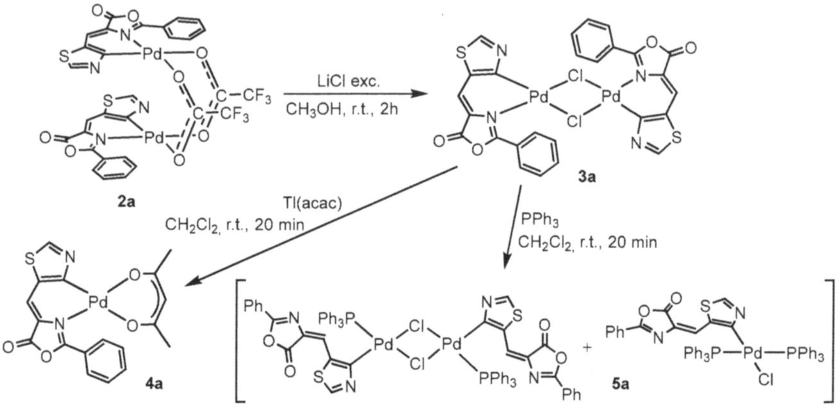

Scheme 3. Synthesis of complexes **3a–5a** from **2a**.

Dimer **3a** is obtained as a brown solid, stable to air and moisture, with a yield of 50%, and is insoluble in common solvents used for NMR characterization. The IR spectrum of **3a** shows a clear absorption at 343 cm^{-1}, assigned to the stretching ν(Pd-Cl), as well as the disappearance of the absorptions assigned to the CF$_3$CO$_2$ ligand in **2a** (1627 cm^{-1} and 1139 cm^{-1}). On the other hand, the HRMS spectrum of **3a** shows a peak at 758.8235 amu, with an isotopic distribution that is in very good agreement with the stoichiometry [Pd$_2$ClC$_{26}$H$_{14}$N$_4$O$_4$S$_2$]$^+$ (758.8219 amu). These observations indicate the metathesis of carboxylate bridges by chloride and the formation of **3a**.

Treatment of **3a** with Tl(acac) (molar ratio 1:2; acac = acetylacetonate) results in the substitution of the chloride ligands with the chelating acac anion, forming the yellow complex **4a** with an 80% yield, as shown in Scheme 3 [24,42]. Elemental analyses of **4a** are consistent with the stoichiometry proposed in Scheme 3. The ^{1}H NMR spectrum confirms the incorporation of the acac ligand as an O,O'-chelate by observing two resonances at 2.06 and 2.04 ppm (relative intensity 3) assigned to the methyl groups, reflecting the asymmetry of the orthopalladated oxazolone ligand, along with a singlet at 5.55 ppm (relative intensity 1) due to the proton of the CH group [42]. The ^{13}C NMR spectrum confirms this asymmetry of the O,O'-acac ligand, as two resonances are observed for the carbonyl groups (188.06 ppm and 185.8 ppm), as well as for the methyl groups (27.36 ppm and 26.61 ppm), while carbon C3 appears at 100.83 ppm in a typical position for the O,O'-chelated ligand [42].

Finally, the cleavage of the chloride bridge system in **3a** with neutral monodentate ligands has been tested. Treatment of a suspension of **3a** in CH$_2$Cl$_2$ with PPh$_3$ (molar ratio 1:2) results in its immediate dissolution, turning the initial brown suspension into a pale-yellow solution. Removal of the solvent yields a solid **5a**, whose ^{31}P spectrum (see Supplementary Materials) shows the presence of two signals (28.64 ppm and 23.24 ppm), suggesting the formation of two different species. In this case, we can rule out the formation of coordination isomers (PPh$_3$-trans-N and PPh$_3$-trans-C), as the reluctance of the PPh$_3$ ligand to coordinate trans to an aromatic C due to the antisymbiotic effect is well established [43]. In closely related works [44], it has been observed that phosphine coordination causes the decoordination of the N atom from the oxazolone, even while maintaining the dinuclear structure. Therefore, it cannot be ruled out that a similar phenomenon is occurring in this case, that is, the coordination of the phosphine and decoordination of the N atom in the dinuclear system, followed by the subsequent rupture with additional phosphine to generate mononuclear systems with two phosphines. By comparing the ^{31}P NMR chemical shifts obtained here with those reported in previous studies, our structural proposal for **5a** is that the obtained mixture corresponds to what is shown in Scheme 3, a dinuclear complex with chloride bridges, an η^1-monodentate oxazolone and a phosphine per Pd, and a mononuclear complex with an η^1-monodentate oxazolone and two *trans* phosphines per Pd [44].

2.4. Photophysical Properties

The optical properties of oxazolones **1a–1c** and complexes **2a** and **4a** were measured in CH$_2$Cl$_2$ solution (10^{-5} M) at 25 °C. The absorption maxima of the oxazolones and their orthopalladated derivatives are listed in Table 1, and the spectra are shown in the Supplementary Materials. The emission spectra, also shown in the Supplementary Materials, were measured using the absorption maxima observed in the UV-visible spectra. Oxazolones **1a–1c** have their maximum (373–386 nm) in the UV A-region, while orthopalladated complexes **2a** and **4a** exhibit their absorbance maximum shifted toward longer wavelengths, in the range of 406–449 nm, corresponding to the blue-ultraviolet region of the UV-Vis spectrum. The presence of absorption bands in this region has been reported for other oxazolones and is attributed to π-π* charge transfer [24,25]. The general bathochromic shift of the absorption maxima from the free ligand to the organometallic complexes has been observed for other palladium-oxazolone complexes and can be interpreted as the palladium contribution to the frontier orbitals. In the case of complex **2a**, two relative absorption maxima are observed, which, as will be discussed later, lead to the same emission.

Table 1. Wavelength values of absorption, excitation and emission maxima (nm).

Compound	$\lambda_{abs.max}$ (nm)	$\lambda_{exc.max}$ (nm)	$\lambda_{emis.max}$ (nm)
1a	378	378	434
1b	386		
1c	373		
2a	425, 449	451	474
4a	406	420	512

Excitation of the compounds at the absorption maxima listed in Table 1 for oxazolone **1a** and complexes **2a** and **4a** results in blue-green emissions (Table 1 and Supplementary Materials). The emission of the complexes is red-shifted compared to that of the free ligand, an effect observed in other palladium complexes with oxazolones [24]. In complex **4a**, a larger Stokes shift is observed, probably reflecting a greater distortion of the molecule in the excited state compared to the ground state. The excitation spectra, recorded at the emission maximum, are consistent with the absorption spectra. Finally, the comparison of the quantum yields measured for **1a** and **2a**, which are both less than 1%, does not allow us to conclude that the presence of the Pd promotes any amplification in this case, as it was observed in other systems [25,26].

3. Materials and Methods

Solvents were obtained from commercial sources and were used without further purification. All reactions were performed without special precautions against air and moisture.

3.1. Synthesis and Characterization of Oxazolones from Heterocycles

The oxazolones **1a–1c** were prepared using the Erlenmeyer–Plöchl method, by reaction of the corresponding hippuric acids and aldehydes in acetic anhydride [27–33].

*Synthesis of (Z)-4-(5-thiazolylmethylene)-2-phenyl-5(4H)-oxazolone **1a***

In a 50 mL round-bottom flask equipped with a reflux system, thiazole-5-carboxaldehyde (1.000 g, 8.57 mmol), hippuric acid (1.567 g, 8.57 mmol), and sodium acetate (0.700 g, 8.57 mmol) were dissolved in acetic anhydride (10 mL, 106.46 mmol). The resulting mixture was heated in an oil bath at 100 °C with stirring for 2 h. After the reaction time, the heating plate was turned off, and the mixture was allowed to cool, resulting in a precipitated solid containing impure oxazolone **1a**. This precipitate was treated with 10 mL of ethanol and stirred for 5 min at room temperature. The insoluble solid was then collected on a filter plate, washed twice with ethanol (2 × 10 mL), and twice with distilled water (2 × 10 mL). The resulting yellow solid was dried under suction and characterized as oxazolone **1a**. Obtained: 2.13 g (yield: 97%). ^{1}H NMR (CDCl$_3$, 400.13 MHz, 298 K): δ = 9.05 (dd, 1H, H$_2$, thiazole), 8.32 (dd, 1H, H$_4$, thiazole), 8.18 (m, 2H, H$_o$, Ph), 7.64 (tt, 1H, H$_p$, Ph, $^3J_{HH}$ = 7.2 Hz, $^4J_{HH}$ = 2.4 Hz), 7.55 (m, 2H, H$_m$, Ph), 7.53 (dd, 1H, H vinyl). ^{13}C NMR (CDCl$_3$, 100.67 MHz, 298 K): δ = 165.82 (C=O), 163.47 (C=N), 159.62 (C$_2$ thiazole), 149.26 (C$_4$, thiazole), 133.57 (C$_p$, Ph), 133.22 (=C), 132.38 (C$_5$, thiazole), 128.89 (C$_m$, Ph), 128.38 (C$_o$, Ph), 125.02 (C$_i$, Ph), 120.97 (=CH, vinyl). HRMS (ESI$^+$) [m/z]: calculated for [C$_{13}$H$_8$N$_2$O$_2$S+K]$^+$ =294.9944; found 294.9967.

*Synthesis of (Z)-4-(4-thiazolylmethylene)-2-phenyl-5(4H)-oxazolone **1b***

Oxazolone **1b** is prepared following the same experimental procedure described for **1a** but using the corresponding aldehyde. Thus, thiazole-4-carbaldehyde (0.500 g, 4.29 mmol), hippuric acid (0.784 g, 4.29 mmol), and sodium acetate (0.350 g, 4.29 mmol) reacted with acetic anhydride (5 mL, 53.23 mmol) under reflux (100 °C, 2 h) to yield oxazolone **1b** as a yellow solid after washing with ethanol (2 × 10 mL) and water (2 × 10 mL). Obtained: 0.43 g (yield: 39%). ^{1}H NMR (CDCl$_3$, 500.13 MHz, 298 K): δ = 8.88 (d, 1H, H$_2$, thiazole, $^4J_{HH}$ = 2 Hz), 8.81 (d, 1H, H$_5$, thiazole, $^4J_{HH}$ = 2 Hz), 8.17 (m, 2H, H$_o$, Ph), 7.62 (tt, 1H, H$_p$, Ph, $^3J_{HH}$ = 7.5 Hz, $^4J_{HH}$ = 1.5 Hz), 7.56 (s, 1H, H vinyl), 7.53 (m, 2H, H$_m$, Ph). ^{13}C NMR (CDCl$_3$, 125.67 MHz, 298 K): δ = 166.58 (C=O), 164.29 (C=N), 152.43 (C$_2$, thiazole), 150.64 (C$_4$, thiazole), 134.06 (=C), 133.56 (C$_p$, Ph), 128.39 (C$_o$, Ph), 128.88 (C$_m$, Ph), 126.07 (C$_5$, thiazole), 125.19 (C$_i$, Ph), 123.81 (=CH, vinyl). HRMS (ESI$^+$) [m/z]: calculated for [C$_{13}$H$_8$N$_2$O$_2$S+K]$^+$ = 294.9944; found 294.9982.

*Synthesis of (Z)-4-(4-(1-acetyl-1H-imidazolylmethylene)-2-phenyl-5(4H)-oxazolone **1c***

Oxazolone **1c** is prepared following the same experimental procedure described for **1a** but using the corresponding aldehyde. Thus, imidazole-4-carbaldehyde (0.500 g, 5.10 mmol), hippuric acid (0.932 g, 5.10 mmol), and sodium acetate (0.420 g, 5.10 mmol) reacted with acetic anhydride (5 mL, 53.23 mmol) under reflux (100 °C, 2 h) to yield oxazolone **1c** as a yellow solid after washing with ethanol (2 × 10 mL) and water (2 × 10 mL). Obtained: 0.81 g (yield: 75%). ^{1}H NMR (CDCl$_3$, 400.13 MHz, 298 K): δ = 8.54 (dd, 1H, H$_2$, imidazole, $^4J_{HH}$ = 1.6 Hz, $^5J_{HH}$ = 0.8 Hz), 8.21 (d, 1H, H$_5$, imidazole, $^4J_{HH}$ = 1.2 Hz), 8.14 (m, 2H, H$_o$, Ph), 7.61 (tt, 1H, H$_p$, Ph, $^3J_{HH}$ = 8 Hz, $^4J_{HH}$ = 2 Hz), 7.52 (m, 2H, H$_m$, Ph), 7.29 (d, 1H, H vinyl, $^5J_{HH}$ = 0.4 Hz), 2.70 (s, 3H, CH$_3$, acetyl). ^{13}C NMR (CDCl$_3$, 100.67 MHz, 298 K): δ = 166.13 (C=O, oxazolone), 165.88 (C=O, acetyl), 163.46 (C=N), 138.27 (C$_4$, imidazole), 136.72 (C$_5$, imidazole), 133.29 (C$_p$, Ph), 133.28 (=C), 128.80 (C$_m$, Ph), 128.25 (C$_o$, Ph), 125.12 (C$_i$, Ph), 123.37 (=CH, vinyl), 121.52 (C$_2$, imidazole), 22.70 (CH$_3$, acetyl). HRMS (ESI$^+$) [m/z]: calculated for [C$_{15}$H$_{11}$N$_3$O$_3$ + Na]$^+$ = 304.0698; found 304.0691.

3.2. Synthesis of Orthopalladated Derivatives **2a–5a** and **2b**

The incorporation of the palladium to the oxazolone ligand through C-H bond activation has been carried out in refluxing trifluoroacetic acid, following the methods previously reported [24–26,36–39].

*Synthesis of orthopalladated complex **2a***

A suspension of Pd(OAc)$_2$ (0.663 g, 2.93 mmol) and **1a** (0.750 g, 2.93 mmol) in CF$_3$CO$_2$H (15 mL) was heated to 70 °C in an oil bath with stirring for 40 min. After the reaction time, the system was allowed to reach room temperature, resulting in the formation of a black suspension. To this suspension, 15 mL of distilled water was added, causing an increase in the precipitated solid. The mixture was stirred at room temperature for 5 min. The black precipitated solid was filtered, washed with 3 portions of distilled water (3 × 10 mL), dried under suction, and characterized as compound **2a** contaminated with black palladium. To recrystallize **2a**, the black solid was treated with CH$_2$Cl$_2$ (about 75 mL), stirred at room temperature for 10 min, and the resulting suspension was filtered through Celite, washing the Celite until the washings were colorless (about 25 mL additional). The resulting intense yellow solution was evaporated to dryness, yielding **2a** as an intense yellow solid. Obtained: 1.513 g (yield: 55%). ^{1}H NMR (CD$_2$Cl$_2$, 500.13 MHz, 298 K): δ = 8.89 (s, 1H, H$_2$ thiazole), 8.53 (m, 2H, H$_o$, Ph), 7.76 (tt, 1H, H$_p$, Ph, $^3J_{HH}$ = 7.5 Hz, $^4J_{HH}$ = 1 Hz), 7.66–7.63 (m, 3H, H$_m$ Ph + H vinyl). ^{13}C NMR (CD$_2$Cl$_2$, 125.67 MHz, 298 K): δ = 170.57 (C=N), 164.50 (C=O), 160.01 (C$_2$, thiazole), 156.7 (Pd-C$_4$, thiazole), 135.95 (C$_p$, Ph), 131.39 (C$_o$, Ph), 128.87 (C$_m$, Ph), 125.03 (=CH, vinyl), 123.46 (C$_5$, thiazole), 122.21 (C$_i$, Ph). The compound has a low solubility, even in CD$_2$Cl$_2$, and the spectra were acquired in long trials using a saturated solution. Despite this, signals due to the quaternary C atoms (=C, C-O and C-F) could not be found in the ^{13}C spectrum due to the low S/N ratio, neither through correlations in the HMBC spectrum. ^{19}F NMR (CDCl$_3$, 470.55 MHz, 298 K): δ = −76.27 (s, CF$_3$). IR (ν, cm^{-1}): 1799 (C=O), 1660, 1571 (O-C=N), 1627 (C-O bridging), 1139 (C-F). Elemental Analysis. Found: C, 38.07; H, 1.40; N, 5.78; S, 6.51. Calculated for C$_{30}$H$_{14}$F$_6$N$_4$O$_8$Pd$_2$S$_2$: C, 37.95; H, 1.49; N, 5.90; S, 6.75.

Synthesis of orthopalladated complex **3a** *[24]*

A suspension of **2a** (0.600 g, 0.63 mmol) in methanol (25 mL) was treated with anhydrous LiCl (0.107 g, 2.52 mmol), and the mixture was stirred at room temperature for 2 h. After the reaction time, the resulting precipitated solid was filtered and washed with 5 mL of cold methanol and 20 mL of diethyl ether. The brown product was dried by suction and characterized as **3a**. Obtained: 0.269 g (yield: 54%). **3a** was found to be completely insoluble in common deuterated solvents, so it could not be characterized by NMR methods. IR (ν, cm^{-1}): 1794 (C=O), 1625 (O-C=N), 343 (Pd-Cl). HRMS (ESI$^+$) [m/z]: calculated for [C$_{26}$H$_{14}$ClN$_4$O$_4$Pd$_2$S$_2$]$^+$ = [M-Cl]$^+$ = 758.8219; found 758.8235.

Synthesis of orthopalladated complex **4a** *[24,42]*

A suspension of **3a** (0.150 g, 0.189 mmol) in CH$_2$Cl$_2$ (15 mL) at room temperature was treated with Tl(acac) (0.115 g, 0.378 mmol). The resulting suspension was stirred at room temperature for 20 min. After this time, the precipitated TlCl was removed by filtration through Celite. The Celite was washed with additional CH$_2$Cl$_2$ (10 mL) and the combined solutions were evaporated to dryness. The resulting residue was treated with 20 mL of cold n-pentane and stirring. The resulting yellow solid (**4a**) was filtered, washed with additional cold n-pentane (2 × 5 mL), dried under suction and characterized as **4a**. Obtained: 0.14 g (yield: 80%). ^{1}H NMR (CDCl$_3$, 500.13 MHz, 298 K): δ = 8.99 (d, 1H, H$_2$, thiazole, $^5J_{HH}$ = 1 Hz), 8.12 (m, 3H, H$_o$, Ph + H vinyl), 7.58 (tt, 1H, H$_p$, Ph, $^3J_{HH}$ = 8 Hz, $^4J_{HH}$ = 2 Hz), 7.50 (m, 2H, H$_m$, Ph), 5.55 (s, 1H, CH, acac), 2.05 (s, 3H, CH$_3$, acac), 2.03 (s, 3H, CH$_3$, acac). ^{13}C NMR (CDCl$_3$, 125.67 MHz, 298 K): δ = 188.06 (C-O, acac), 185.80 (C-O, acac), 170.66 (C$_4$, thiazole), 166.74 (C=O), 161.44 (C=N), 158.42 (C$_2$, thiazole), 132.92 (C$_p$, Ph), 130.55 (=C), 129.98 (C$_5$, thiazole), 128.81 (C$_m$, Ph), 128.07 (C$_o$, Ph), 127.86 (=CH, vinilo), 125.60 (C$_i$, Ph), 100.83 (CH, acac), 27.36 (CH$_3$, acac), 26.61 (CH$_3$, acac). Elemental Analysis. Found: C, 46.66; H, 3.37; N, 5.73; S, 7.09. Calculated for C$_{18}$H$_{14}$N$_2$O$_4$PdS: C, 46.92; H, 3.06; N, 6.08; S, 6.96.

Reactivity of **3a** *with PPh$_3$* (**5a**)

A suspension of **3a** (0.150 g, 0.19 mmol) in CH$_2$Cl$_2$ (15 mL) at room temperature was treated with PPh$_3$ (0.100 g, 0.38 mmol). The resulting suspension was stirred at room temperature for 20 min, during which time the initial suspension dissolved. After this period, any remaining insoluble solid was removed by filtration. The obtained solution was evaporated to dryness, and the residue was treated with 20 mL of cold *n*-pentane with stirring. The resulting pale yellow solid was filtered, washed with two additional 5 mL portions of cold *n*-pentane, and dried by suction. The resulting solid was characterized as the mixture **5a** (see text). Obtained: 0.06 g. ^{31}P NMR (CDCl$_3$, 202.46 MHz, 298 K): δ = 28.64 ppm (s, 1P, PPh$_3$), 23.24 ppm (s, 2P, PPh$_3$).

*Synthesis of orthopalladated complex co-crystallized with [Pdpy$_4$](CF$_3$CO$_2$)$_2$ (**2b**)*

A suspension of Pd(OAc)$_2$ (0.137 g, 0.61 mmol) and **1b** (0.157 g, 0.61 mmol) in CF$_3$CO$_2$H (10 mL) was heated to 70 °C in an oil bath with stirring for 40 min. After this time, the reaction was cooled to room temperature, resulting in the formation of a deep red suspension. To this suspension, 15 mL of distilled water was added, resulting in an increase in the amount of precipitated solid. The precipitated solid was filtered, washed with additional portions of distilled water (3 × 10 mL), one portion of acetone (10 mL), one portion of CH$_2$Cl$_2$ (10 mL), and dried by suction. The solid thus generated (0.130 g) showed to be totally insoluble in the usual deuterated solvents. However, a suspension of this insoluble solid (10 mg) in CD$_2$Cl$_2$ (0.6 mL) reacted with 1 drop of py-d$_5$ producing the instantaneous solution of the initial suspension and the obtention of a clear yellow solution, containing **2b** and other species, which was partially characterized by ^{1}H NMR. Slow evaporation of this solution gave a small crop of crystals of (**2b**), which was characterized as the co-crystallization of [Pd(N^N^C)(py)] and [(Pdpy$_4$)(CF$_3$CO$_2$)$_2$]$_{0.5}$(CD$_2$Cl$_2$)$_{0.8}$.

3.3. Spectroscopic and Analytical Methods

The ^{1}H, ^{13}C and ^{19}F NMR spectra of the isolated products were recorded in CDCl$_3$ or CD$_2$Cl$_2$ solutions at 25 °C on Bruker AV300, AV400, or Bruker AV500 spectrometers (δ in ppm, *J* in Hz) at ^{1}H operating frequencies of 300.13, 400.13, and 500.13 MHz, respectively. The ^{1}H and ^{13}C NMR spectra were referenced using the solvent signal as the internal standard, while ^{19}F NMR spectra were referenced to CFCl$_3$. The ^{13}C NMR peaks were identified using standard ^{1}H–^{13}C edited-HSQC and ^{1}H–^{13}C HMBC 2D-experiments. In both cases 4K points in t$_2$ using spectral widths of 10 ppm (^{1}H) and 200 ppm (^{13}C) were used, with averaged values of the coupling constants $^1J_{CH}$ = 145 Hz and long-range $^nJ_{CH}$ = 10 Hz. Typically, 128 t$_1$ experiments were recorded and zero-filled to 2K. For each t$_1$ value 8 (HSQC) or 32 (HMBC) scans were signal-averaged using a recycle delay of 1 s. High-resolution mass spectra-ESI (HRMS-ESI) were recorded using either a Bruker MicroToF-Q™ system equipped with an API-ESI source and a Q-ToF mass analyzer, or a TIMS-TOF system, both allowing a maximum error in the measurement of 5 ppm. Acetonitrile was used as solvent. For all types of MS measurements, samples were introduced in a continuous flow of 0.2 mL/min and nitrogen served both as the nebulizer gas and the dry gas. Absorption spectra were measured on a Thermo Scientific Evolution 600 BB spectrophotometer. The steady-state excitation–emission spectra were measured on a FluoTime300 PicoQuant spectrometer. All measurements were carried out at room temperature on solutions of 10^{-5} M concentration using quartz cuvettes of 1 cm path length. All measurements were carried out at room temperature on solutions of 10^{-5} M concentration using quartz cuvettes of 1 cm path length. Quantum yields of 5×10^{-4} M solutions of **1a** and **2a** in CH$_2$Cl$_2$ were measured by the absolute method using a Hamamatsu Quantaurus-QY C11347 compact one-box absolute quantum yield measurement system. Through studies carried out for different substances using both the absolute method and the comparative one, the relative uncertainty for the absolute method has been determined as less than 6% [45].

3.4. X-Ray Crystallography

Crystals of compounds **1c** were obtained by slow diffusion of *n*-pentane into a solution of the crude compound in CH$_2$Cl$_2$ at −18 °C over the course of one week, while crystals of **2b** ([Pd(N^N^C)(py)] co-crystallized with [Pd(py)$_4$](CF$_3$CO$_2$)$_2$) were obtained by slow evaporation of a CD$_2$Cl$_2$ solution containing both compounds (see text). **Data collection.** The selected crystal of **1c** was placed on a micromount coated with protecting perfluoropolyether oil and cooled to 100 K with an open-flow nitrogen gas. X-ray diffrac-

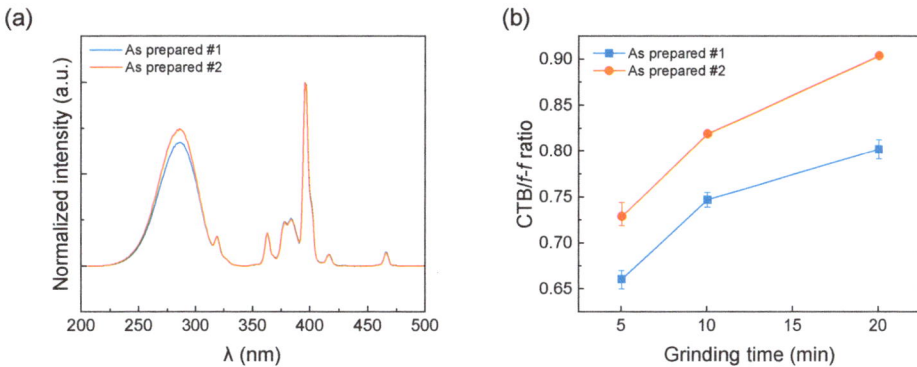

Figure 1. (**a**) Photoluminescence excitation (PLE) spectra of two different Si-substituted CaYAlO$_4$:Eu (CYASO:Eu) samples prepared under the same conditions nominally. (**b**) The ratios of charge transfer band (CTB) to f-f transition for the different grinding times.

2.2. Ball Milling

The sample preparation using a solid-state reaction may include ball milling. The ball milling process is used for grinding and mixing materials and often to reduce the particle size. We attempted to reduce the particle size and induce surface defects in Al$_2$O$_3$, one of the starting materials, to see their effects on the photoluminescent response. We used a planetary ball mill machine with a Teflon container (i.e., milling jar) for the ball milling process. Initially, we used zirconia (ZrO$_2$) balls, which are commonly used. Compared to the pristine Al$_2$O$_3$ powder, the widths of the XRD peaks become broader after a ball milling process for 3 h (Figure 2a), indicating reduced particle size. However, additional peaks corresponding to the ZrO$_2$ phase emerge. The Mors hardness values of Al$_2$O$_3$ (9.0) and ZrO$_2$ (6.5) suggest possible scraps from the ZrO$_2$ ball. The impurity phase, such as Fe or Ni, due to the ball milling process with stainless steel balls has been investigated [21]. However, detecting insignificant contamination from ceramic balls might be very challenging, especially after the entire solid-state reaction process. Hence, careful consideration should be given when selecting materials for ball milling processes.

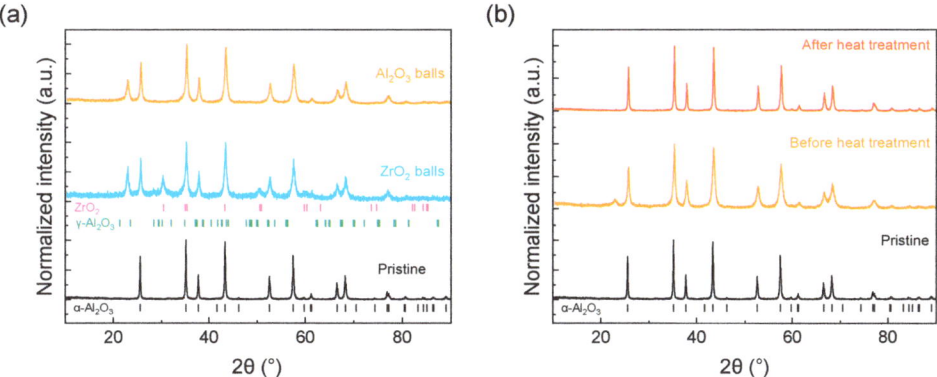

Figure 2. (**a**) X-ray diffraction (XRD) patterns of Al$_2$O$_3$ powders after the planetary ball milling using ZrO$_2$ or Al$_2$O$_3$ balls. The color bars indicate the reference diffraction peak positions: α-Al$_2$O$_3$ (JCPDS No. 42-1468), γ-Al$_2$O$_3$ (JCPDS No. 52-0803), and ZrO$_2$ (JCPDS No. 50-1089). (**b**) XRD patterns of the ball-milled Al$_2$O$_3$ powder before and after heat treatment at 800 °C for 1 h. The XRD pattern of the pristine Al$_2$O$_3$ powder is included for comparison.

Using Al_2O_3 balls could resolve this issue. However, attention must also be paid to scraps that peel off from the container. While the ball milling process with Al_2O_3 balls excludes the ZrO_2 contamination, an additional peak at ≈22.93° persists irrespective of the kind of balls (Figure 2a). This peak is unrelated to ZrO_2 or both α- and γ-Al_2O_3 phases. Heat treatment at 800 °C eliminates this peak (Figure 2b), implying that the phase might be organic scraps from the Teflon container. Since solid-state reaction includes a high-temperature heating process, one may think that the organic scraps are not an issue. However, it can create a significant issue related to off-stoichiometry if the ball milling process is used for precursors individually.

As previously mentioned, we aimed to reduce the particle size and induce surface defects in Al_2O_3, one of the starting materials, to examine its influence on the photoluminescent properties of the final compound, CYASO:Eu. We performed the ball milling process on Al_2O_3 for 1 to 10 h and checked the change in the width of XRD peaks (Figure 3a). The peak width becomes broader after 1 h of ball milling, indicative of the reduced particle size. However, additional milling does not significantly alter the peak width. Using the Scherrer formula, we estimated the particle sizes from the XRD peak width. The particle size reduces from 40.0 nm to 13.5 nm after 1 h of ball milling (the inset of Figure 3a).

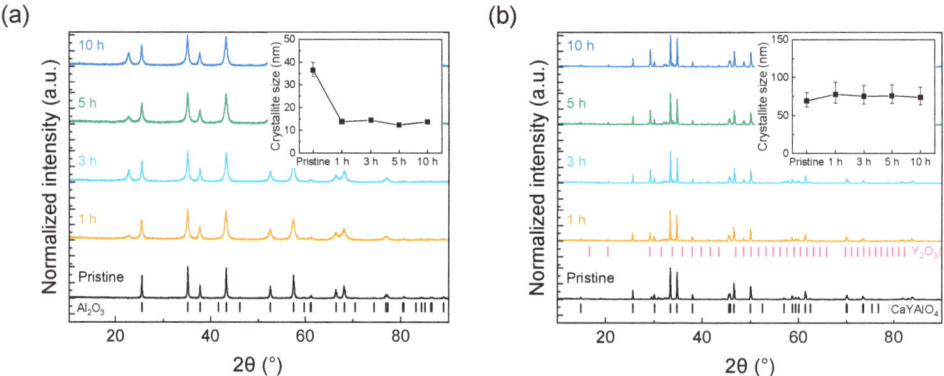

Figure 3. (**a**) X-ray diffraction (XRD) patterns of Al_2O_3 powders after ball milling for different times (0, 1, 3, 5, and 10 h). The color bars indicate the reference diffraction peak positions: α-Al_2O_3 (JCPDS No. 42-1468). The inset represents the particle size evaluated by the Scherrer equation. (**b**) XRD patterns of Si-substituted $CaYAlO_4$:Eu (CYASO:Eu) samples prepared with the Al_2O_3 powders subjected to the different ball milling times. The color bars indicate the reference diffraction peak positions: $CaYAlO_4$ (JCPDS No. 81-0742) and Y_2O_3 (JCPDS No. 41-1105). The inset shows the particle size evaluated by the Scherrer equation applied for the (103) peaks of the $CaYAlO_4$ phase.

We noticed that the total mass of powders increases after the ball milling compared to the mass of initially loaded Al_2O_3 powder, and the total mass was proportional to the milling time. We attribute the increased mass after the ball milling to the peeled-off scraps from the Teflon container. A problem may arise when we take and use some portion of the ball-milled powder for subsequent processes, and if we need to mix powders by considering the stoichiometry of a final compound, we are not able to determine the mass of ball-milled powder for mixing because chemical formula or mixture ratio of ball-milled powder is unknown. Such an issue becomes evident in the subsequent experimental results. While the overall mass of the powder increased after milling, we assumed that we did not recognize the mass increase, and we took only the required mass of Al_2O_3 for the stoichiometric CYASO synthesis. As milling time increases, so does mass; hence, taking the equal masses of the precursors prepared under different milling times causes adding different moles of a specific element, potentially leading to defective samples. In other words, although one intended to systematically prepare samples using different Al_2O_3

tion data of **1c** was collected at XALOC beamline [46] of Alba-CELLS synchrotron (Spain) (λ = 0.72940 Å) measuring 360 phi scans with two different κ angle values, on a single axis MD2 Maatel diffractometer and a PILATUS 6M detector. Diffraction images were integrated using SAINT [47], and the integrated intensities were scaled and corrected for absorption using SADABS program [48], included in APEX2 package. The selected crystal of **2b** was placed randomly at the end of a glass fiber and covered with perfluorinated oil (magic oil). Data collection was performed at 295 K (**2b**) using an Oxford Diffraction Xcalibur diffractometer with Mo-Kα radiation (graphite monochromator, λ = 0.71073 Å). A hemisphere of data was collected based on scans of ω and φ angles, and the diffraction images were integrated and corrected for absorption using CrysAlis Pro software (https://rigaku.com/products/crystallography/x-ray-diffraction/crysalispro) [49]. **Structure solution and refinement.** The structures were solved using direct methods [50,51]. Refinement of all non-hydrogen atoms was carried out with anisotropic displacement parameters. Hydrogen atoms of **1c** were included in the model in observed positions and freely refined. Hydrogen atoms of **2b** were placed in positions calculated mathematically as riding atoms on the atoms to which they are bonded (C, N, O), except those noted below, and were assigned an isotropic displacement parameter equal to 1.2 or 1.5 times the equivalent isotropic displacement parameter of the bonded atom. The structures were refined with respect to F_o^2, and all reflections were used in the least-squares calculations [51,52]. The resolution of the crystal structure of **2b** shows that the asymmetric part of the unit cell contains one molecule of the orthometalated complex [Pd(N^N^C)(py)], half of a [Pdpy4]$^{2+}$, one (CF$_3$CO$_2$)$^-$, and 0.4 of dichloromethane. The CF$_3$ group has a rotational disorder in the F atoms, modeled with occupancies of 0.8/0.2. Constraints are applied to the C-F distances (SAME) and the thermal parameters of the F atoms. The position of the hydrogen atom of the OH fragment is found in the electron density maps. It is refined using a free isotropic parameter with a fixed O-H distance of 0.82 Å. There is a hydrogen bond between this OH group and an oxygen atom of the trifluoroacetate. CCDC **2384620** (**2b**) and **2384671** (**1c**) contain the supplementary crystallographic data. These data can be obtained free of charge from the Cambridge Crystallographic Data Centre via www.ccdc.cam.ac.uk/structures/.

4. Conclusions

In summary, the synthesis of (Z)-4-hetarylidene-5(4H)-oxazolones **1** for various heterocycles (thiazole, imidazole) has been carried out, and it has been found that they are not fluorescent species. It has also been demonstrated that the incorporation of Pd by C-H activation into the heterocycle located at the 4-hetarylidene position in various organometallic complexes **2–5** does not lead, in these cases, to a clearly detectable fluorescence amplification due to the suppression of the hula-twist deactivation pathway. Unexpectedly, it has been observed in thiazole derivatives that the different location of the vinyl group –C(H)=C (5 vs. 4) relative to the heteroatoms of the thiazole (N, S) affects the position of C-H activation and subsequent orthopalladation. Thus, when the vinyl group is in position 5, the thiazole heterocycle is orthopalladated at position 4, while when the vinyl group is in position 4, N-coordination of the thiazole occurs, and the unprecedented activation of the phenyl ring in position 2 of the oxazolone is produced.

Supplementary Materials: The following supporting information can be downloaded at https://www.mdpi.com/article/10.3390/inorganics12100271/s1. Copies of ^{1}H and ^{13}C NMR spectra of all new species; UV-VIS Absorption and emission spectra of all new species; Table S1: Crystal data and structure refinement for **1c**; Table S2: Bond distances (Å) and angles (°) for **1c**; Table S3: Crystal data and structure refinement for **2b**; Table S4. Bond distances (Å) and angles (°) for **2b**; ORTEP drawings of the contents of the asymmetric part of the unit cell of **2b**.

Author Contributions: Conceptualization, D.D. and E.P.U.; methodology, D.D. and M.M.; validation, M.M., D.D. and E.P.U.; formal analysis, M.M., D.D., O.C., P.G.-O., F.L. and A.M.; investigation, M.M., D.D., O.C., P.G.-O., F.L. and A.M.; resources, E.P.U.; data curation, M.M., D.D., O.C., P.G.-O., F.L. and A.M.; writing—original draft preparation, M.M. and E.P.U.; writing—review and editing, all authors;

supervision, E.P.U.; project administration, E.P.U.; funding acquisition, E.P.U. All authors have read and agreed to the published version of the manuscript.

Funding: Spanish Government (grants PID2019-106394GB-I00 and PID2022-136861NB-I00, funded by MCIN/AEI/10.13039/501100011033, and grant PID2021-122869NB-I00 funded by MICIU/AEI/10.13039/501100011033 and ERDF/EU.) and Gobierno de Aragón-FSE (Spain, research groups Química Inorgánica y de los Compuestos Organometálicos E17_23R and Química de Oro y Plata E07_23R).

Data Availability Statement: Additional data are contained within the article and Supplementary Materials.

Acknowledgments: E.P.U. and D.D. thank the Spanish Government (grant PID2019-106394GB-I00, funded by MCIN/AEI/10.13039/501100011033); A.M. also thanks the Spanish Government (grant PID2021-122869NB-I00 funded by MICIU/AEI/10.13039/501100011033 and ERDF/EU); E.P.U., A.M. and D.D. thank Gobierno de Aragón-FSE (Spain, research groups Química Inorgánica y de los Compuestos Organometálicos E17_23R) for funding. D. Dalmau thanks Gobierno de Aragón-FSE for a PhD fellowship. Some X-ray diffraction experiments were performed at XALOC beamline at ALBA Synchrotron with the collaboration of ALBA staff.

Conflicts of Interest: The authors declare no conflicts of interest.

References

1. Valeur, B.; Berberan-Santos, M.N. *Molecular Fluorescence, Principles and Applications*, 2nd ed.; Wiley-VCH: Weinheim, Germany, 2013.
2. Edgar, A. Luminescent Materials. In *Springer Handbook of Electronic and Photonic Materials*; Kasap, S., Capper, P., Eds.; Springer Handbooks; Springer: Cham, Switzerland, 2017.
3. Kitai, A. *Luminescent Materials and Applications*, 1st ed.; Wiley-VCH: Weinheim, Germany, 2008.
4. Kandarakis, I.S. Luminescence in medical image science. *J. Lumin.* **2016**, *169*, 553–558. [CrossRef]
5. Ronda, C. Challenges in Application of Luminescent Materials, a Tutorial Overview. *Prog. Electromagn. Res.* **2014**, *147*, 81–93. [CrossRef]
6. Prodi, L.; Montalti, M.; Zaccheroni, N. Luminescence Applied in Sensor Science. *Top. Curr. Chem.* **2011**, *300*, 1–217.
7. Feldmann, C.; Jüstel, T.; Ronda, C.R.; Schmidt, P.J. Inorganic Luminescent Materials: 100 Years of Research and Application. *Adv. Funct. Mater.* **2003**, *13*, 511–516. [CrossRef]
8. Shimomura, O. Discovery of green fluorescent protein (GFP) (Nobel Lecture). *Angew. Chem. Int. Ed.* **2009**, *48*, 5590–5602. [CrossRef]
9. Chalfie, M. GFP: Lighting Up Life (Nobel Lecture). *Angew. Chem. Int. Ed.* **2009**, *48*, 5603–5611. [CrossRef]
10. Tsien, R.Y. Constructing and Exploiting the Fluorescent Protein Paintbox (Nobel Lecture). *Angew. Chem. Int. Ed.* **2009**, *48*, 5612–5626. [CrossRef]
11. Remington, S.J. Fluorescent proteins: Maturation, photochemistry and photophysics. *Curr. Opin. Struct. Biol.* **2006**, *16*, 714–721. [CrossRef]
12. Pakhomov, A.A.; Martynov, V.I. GFP Family: Structural Insights into Spectral Tuning. *Chem. Biol.* **2008**, *15*, 755–764. [CrossRef]
13. Wachter, R.M. Chromogenic Cross-Link Formation in Green Fluorescent Protein. *Acc. Chem. Res.* **2007**, *40*, 120–127. [CrossRef]
14. Mizuno, H.; Mal, T.K.; Tong, K.I.; Ando, R.; Furuta, T.; Ikura, M.; Miyawaki, A. Photo-Induced Peptide Cleavage in the Green-to-Red Conversion of a Fluorescent Protein. *Mol. Cell.* **2003**, *12*, 1051–1058.
15. Collado, S.; Pueyo, A.; Baudequin, C.; Bischoff, L.; Jiménez, A.I.; Cativiela, C.; Hourau, C.; Urriolabeitia, E.P. Orthopalladation of GFP-Like Fluorophores Through C–H Bond Activation: Scope and Photophysical Properties. *Eur. J. Org. Chem.* **2018**, *2018*, 6158–6166. [CrossRef]
16. Ivashkin, P.E.; Yampolsky, I.V.; Lukyanov, K.A. Synthesis and properties of chromophores of fluorescent proteins. *Russ. J. Bioorg. Chem.* **2009**, *35*, 652–669. [CrossRef] [PubMed]
17. Follenius-Wund, A.; Bourotte, M.; Schmitt, M.; Iyice, F.; Lami, H.; Bourguignon, J.J.; Haiech, J.; Pigault, C. Fluorescent derivatives of the GFP chromophore give a new insight into the GFP fluorescence process. *Biophys. J.* **2003**, *85*, 1839–1850.
18. Rajbongshi, B.K.; Sen, P.; Ramanathan, G. Twisted intramolecular charge transfer in a model green fluorescent protein luminophore analog. *Chem. Phys. Lett.* **2010**, *494*, 295–300. [CrossRef]
19. Martin, M.E.; Negri, F.; Olivucci, M. Origin, Nature, and Fate of the Fluorescent State of the Green Fluorescent Protein Chromophore at the CASPT2//CASSCF Resolution. *J. Am. Chem. Soc.* **2004**, *126*, 5452–5464. [CrossRef]
20. Altoe', P.; Bernardi, F.; Garavelli, M.; Orlandi, G.; Negri, F. Solvent effects on the vibrational activity and photodynamics of the green fluorescent protein chromophore: A quantum-chemical study. *J. Am. Chem. Soc.* **2005**, *127*, 3952–3963. [CrossRef]
21. Megley, C.M.; Dickson, L.A.; Maddalo, S.L.; Chandler, G.J.; Zimmer, M. Photophysics and dihedral freedom of the chromophore in yellow, blue, and green fluorescent protein. *J. Phys. Chem. B* **2009**, *113*, 302–308. [CrossRef]
22. Wu, L.; Burgess, K. Syntheses of Highly Fluorescent GFP-Chromophore Analogues. *J. Am. Chem. Soc.* **2008**, *130*, 4089–4096. [CrossRef]

23. Baranov, M.S.; Lukyanov, K.A.; Borissova, A.O.; Shamir, J.; Kosenkov, D.; Slipchenko, L.V.; Tolbert, L.M.; Yampolsky, I.V.; Solntsev, K.M. Conformationally Locked Chromophores as Models of Excited-State Proton Transfer in Fluorescent Proteins. *J. Am. Chem. Soc.* **2012**, *134*, 6025–6032. [CrossRef]
24. Laga, E.; Dalmau, D.; Arregui, S.; Crespo, O.; Jiménez, A.I.; Pop, A.; Silvestru, C.; Urriolabeitia, E.P. Fluorescent Orthopalladated Complexes of 4-Aryliden-5(4*H*)-oxazolones from the Kaede Protein: Synthesis and Characterization. *Molecules* **2021**, *26*, 1238. [CrossRef] [PubMed]
25. Dalmau, D.; Crespo, O.; Matxain, J.M.; Urriolabeitia, E.P. Fluorescence Amplification of Unsaturated Oxazolones Using Palladium: Photophysical and Computational Studies. *Inorg. Chem.* **2023**, *62*, 9792–9806. [CrossRef] [PubMed]
26. Dumitras, D.; Dalmau, D.; García-Orduña, P.; Pop, A.; Silvestru, A.; Urriolabeitia, E.P. Orthopalladated imidazolones and thiazolones: Synthesis, photophysical properties and photochemical reactivity. *Dalton Trans.* **2024**, *53*, 8948–8967. [CrossRef] [PubMed]
27. Erlenmeyer, E. Ueber die Condensation der Hippursäure mit Phtalsäureanhydrid und mit Benzaldehyd. *Justus Liebigs Ann. Der Chem.* **1893**, *275*, 1–8.
28. Plöchl, J. Ueber Phenylglycidasäure (Phenyloxacrylsäure). *Chem. Ber.* **1883**, *16*, 2815. [CrossRef]
29. Plöchl, J. Ueber einige Derivate der Benzoylimidozimmtsäure. *Chem. Ber.* **1884**, *17*, 1616. [CrossRef]
30. Carter, H.E. Azlactones. *Org. React.* **1946**, *3*, 198–237.
31. Filler, R. *Advances in Heterocyclic Chemistry*; Katritzky, A.R., Ed.; Academic Press: New York, NY, USA, 1954; Volume 4, p. 75.
32. Rao, Y.S.; Filler, R. Geometric Isomers of 2-Aryl(Aralkyl)-4-arylidene(alkylidene)-5(4*H*)-oxazolones. *Synthesis* **1975**, *12*, 749–764. [CrossRef]
33. Rao, Y.S.; Filler, R. Oxazoles. In *The Chemistry of Heterocyclic Compounds*; Turchi, I.J., Ed.; John Wiley & Sons, Inc.: New York, NY, USA, 1986; Volume 3, pp. 363–691.
34. Prokof'ev, E.P.; Karpeiskaya, E.I. The Proton Coupled ^{13}C NMR Direct Determination of Z-, E-Configuration of 4-Benzyliden-2-Phenyl(Methyl)- Δ^2-Oxazolin-5-Ones and Products of Their Solvolysis. *Tetrahedron Lett.* **1979**, *20*, 737–740. [CrossRef]
35. Vogeli, U.; von Phillipsborn, W. Vicinal C,H Spin Coupling in Substituted Alkenes. Stereochemical Significance and Structural Effects. *Org. Magn. Reson.* **1975**, *7*, 617–627. [CrossRef]
36. Roiban, G.D.; Serrano, E.; Soler, T.; Aullón, G.; Grosu, I.; Cativiela, C.; Martínez, M.; Urriolabeitia, E.P. Regioselective Orthopalladation of (Z)-2-Aryl-4-Arylidene-5(4*H*)-Oxazolones: Scope, Kinetico-Mechanistic, and Density Functional Theory Studies of the C–H Bond Activation. *Inorg. Chem.* **2011**, *50*, 8132–8143. [CrossRef] [PubMed]
37. Roiban, D.; Serrano, E.; Soler, T.; Grosu, I.; Cativiela, C.; Urriolabeitia, E.P. Unexpected [2 + 2] C–C bond coupling due to photocycloaddition on orthopalladated (Z)-2-aryl-4-arylidene-5(4*H*)-oxazolones. *Chem. Commun.* **2009**, 4681–4683. [CrossRef] [PubMed]
38. Serrano, E.; Juan, A.; García-Montero, A.; Soler, T.; Jiménez-Márquez, F.; Cativiela, C.; Gomez, M.V.; Urriolabeitia, E.P. Stereoselective Synthesis of 1,3-Diaminotruxillic Acid Derivatives: An Advantageous Combination of C–H–ortho–Palladation and On-Flow [2 + 2]–Photocycloaddition in Microreactors. *Chem. Eur. J.* **2016**, *22*, 144–152. [CrossRef] [PubMed]
39. Carrera, C.; Denisi, A.; Cativiela, C.; Urriolabeitia, E.P. Functionalized 1,3-Diaminotruxillic Acids by Pd-Mediated C–H Activation and [2 + 2]–Photocycloaddition of 5(4*H*)–Oxazolones. *Eur. J. Inorg. Chem.* **2019**, *2019*, 3481–3489. [CrossRef]
40. Guy Orpen, A.; Brammer, L.; Allen, F.H.; Kennard, O.; Watson, D.G.; Taylor, R. Supplement. Tables of bond lengths determined by X-ray and neutron diffraction. Part 2. Organometallic compounds and co-ordination complexes of the d- and f-block metals. *J. Chem. Soc. Dalton Trans.* **1989**, S1–S83.
41. Sierra, S.; Gomez, M.V.; Jiménez, A.I.; Pop, A.; Silvestru, C.; Marín, M.L.; Boscá, F.; Sastre, G.; Gómez-Bengoa, E.; Urriolabeitia, E.P. Stereoselective, Ruthenium-Photocatalyzed Synthesis of 1,2-Diaminotruxinic Bis-amino Acids from 4-Arylidene-5(4*H*)-oxazolones. *J. Org. Chem.* **2022**, *87*, 3529–3545. [CrossRef]
42. Forniés, J.; Martínez, F.; Navarro, R.; Urriolabeitia, E.P. Synthesis and reactivity of acetylacetonato-Cγ complexes of M^{II} (M = Pd or Pt): X-ray crystal structure of [Pd(C$_6$F$_5$)(OOCPh)(bipy)]. *J. Organomet. Chem.* **1995**, *495*, 185–194. [CrossRef]
43. Pearson, R. Antisymbiosis and the trans effect. *Inorg. Chem.* **1973**, *12*, 712–713. [CrossRef]
44. Roiban, D.; Serrano, E.; Soler, T.; Contel, M.; Grosu, I.; Cativiela, C.; Urriolabeitia, E.P. Ortho-Palladation of (Z)-2-Aryl-4-Arylidene-5(4*H*)-Oxazolones. Structure and Functionalization. *Organometallics* **2010**, *29*, 1428–1435. [CrossRef]
45. Würth, C.; Grabolle, M.; Pauli, J.; Spieles, M.; Resch-Genger, U. Comparison of Methods and Achievable Uncertainties for the Relative and Absolute measurement of Photoluminescence Quantum Yields. *Anal. Chem.* **2011**, *83*, 3431–3439. [CrossRef]
46. Juanhuix, J.; Gil-Ortiz, F.; Cuní, G.; Colldelram, C.; Nicolás, J.; Lidón, J.; Boter, E.; Ruget, C.; Ferrer, S.; Benach, J. Developments in optics and performance at BL13-XALOC, the macromolecular crystallography beamline at the Alba synchrotron. *J. Synchr. Rad.* **2014**, *21*, 679–689. [CrossRef] [PubMed]
47. *SAINT Software Reference Manuals, Version V8.40B in APEX4*; Bruker Analytical Xray Systems, Inc.: Madison, WI, USA, 2016.
48. Krause, L.; Herbst-Irmer, R.; Sheldrick, G.M.; Stalke, D.J. SADABS 2016/2. *Appl. Cryst.* **2015**, *48*, 3–10. [CrossRef] [PubMed]
49. *CrysAlis RED, CCD Camera Data Reduction Program*; Rigaku Oxford Diffraction: Oxford, UK, 2019.
50. Sheldrick, G.M. SHELXS-86, Phase annealing in SHELX-90: Direct methods for larger structures. *Acta Crystallogr.* **1990**, *A46*, 467. [CrossRef]

51. Sheldrick, G.M. *SHELXS 97 and SHELXL 97, Program for Crystal Structure Solution and Refinement*; University of Göttingen: Göttingen, Germany, 1997.
52. Sheldrick, G.M. Crystal structure refinement with SHELXL. *Acta Cryst. Sect. C Struct. Chem.* **2015**, *C71*, 3.

Disclaimer/Publisher's Note: The statements, opinions and data contained in all publications are solely those of the individual author(s) and contributor(s) and not of MDPI and/or the editor(s). MDPI and/or the editor(s) disclaim responsibility for any injury to people or property resulting from any ideas, methods, instructions or products referred to in the content.

Article

Effect of Synthesis Conditions on the Photoluminescent Properties of Si-Substituted CaYAlO$_4$:Eu: Sources of Experimental Errors in Solid-State Synthesis

Ju Hyun Oh [1], Yookyoung Lee [1], Jihee Kim [2], Woo Tae Hong [3], Hyun Kyoung Yang [4], Mijeong Kang [2] and Seunghun Lee [1,*]

[1] Department of Physics, Pukyong National University, Busan 48513, Republic of Korea; juhyun@pknu.ac.kr (J.H.O.); lyk9746@pukyong.ac.kr (Y.L.)
[2] Department of Optics and Mechatronics Engineering, Pusan National University, Busan 46241, Republic of Korea; alpha11@pusan.ac.kr (J.K.); mkang@pusan.ac.kr (M.K.)
[3] Marine-Bionics Convergence Technology Center, Pukyong National University, Busan 48513, Republic of Korea; hwt525@pknu.ac.kr
[4] Department of Electrical, Electronics and Software Engineering, Pukyong National University, Busan 48547, Republic of Korea; hkyang@pknu.ac.kr
* Correspondence: seunghun@pknu.ac.kr

Abstract: To improve the luminescent efficiency of and to design the color spectrum of phosphors, the comprehensive understanding of the correlation between physical parameters and luminescent properties is imperative, necessitating systematic experimental studies. However, unintentional variations across individually prepared samples impede the thorough investigation of the correlation. In this study, we investigate the possible sources of unintentional variation in the photoluminescence properties of phosphors during sample preparation using a solid-state reaction, explicitly focusing on the ball milling process. Based on the quantitative features of the photoluminescent properties and their associated statistical errors, we explore the impact of unintentional variation alongside intended systematic variation, highlighting its potential to obscure meaningful trends.

Keywords: phosphors; solid-state reaction; ball milling; CaYAlO$_4$; photoluminescence

Citation: Oh, J.H.; Lee, Y.; Kim, J.; Hong, W.T.; Yang, H.K.; Kang, M.; Lee, S. Effect of Synthesis Conditions on the Photoluminescent Properties of Si-Substituted CaYAlO$_4$:Eu: Sources of Experimental Errors in Solid-State Synthesis. *Inorganics* **2024**, *12*, 150. https://doi.org/10.3390/inorganics12060150

Academic Editors: Maurizio Peruzzini and Binbin Chen

Received: 10 May 2024
Revised: 27 May 2024
Accepted: 28 May 2024
Published: 30 May 2024

Copyright: © 2024 by the authors. Licensee MDPI, Basel, Switzerland. This article is an open access article distributed under the terms and conditions of the Creative Commons Attribution (CC BY) license (https:// creativecommons.org/licenses/by/ 4.0/).

1. Introduction

Phosphor-converted white-light-emitting diodes (pc-wLEDs) have emerged as the next-generation lighting technology, promising enhanced efficiency and longevity compared to traditional lighting sources [1–3]. The most common method for WLED production involves utilizing blue LEDs as the primary light source and incorporating yellow phosphors to achieve white light emission. While this approach offers the advantage of simplicity in obtaining white light, it has several issues, including a high color temperature [4,5]. To address the issues in the conventional method, alternative approaches have been explored, including the integration of blue, green, and red phosphors [6–8] or the development of single-component white phosphors containing multiple luminescent activators [9,10]. To improve luminescence efficiency as well as to achieve precise control over emission characteristics, the thorough understanding of the correlation between physical parameters and the luminescent properties of a phosphor is essential.

Beyond simply examining emission or excitation wavelengths, it is required to scrutinize subtle changes in both photoluminescent excitation and emission spectra, such as the peak intensity ratio of the charge transfer band (CTB) to f-f transition, Judd–Ofelt parameters, and asymmetry ratios (electric dipole transition to magnetic dipole transition) [11–15]. An intriguing aspect stems from the inconsistency observed in the literature regarding the ratio of CTB to f-f transition in PLE spectra of Eu-doped oxide phosphors [16]. This inconsistency, lacking a clear explanation, hints at unintentional variation inherent in the synthesis process, which likely affects the ratio of CTB to f-f transition. Oxygen vacancy is one of the most common and inevitable defects in oxides; thus, oxygen vacancy has been attributed to this ratio, yet a comprehensive understanding of its origin remains elusive [17,18]. A thorough grasp of the factors correlated with an experimental feature and a keen awareness of statistical error, such as standard deviation, are essential for drawing accurate conclusions.

In this study, we discuss the potential sources of unintentional change in the photoluminescence properties of phosphors prepared via solid-state reaction. By investigating the change in the structural and photoluminescent properties of Si-substituted $CaYAlO_4$:Eu phosphors under different sample preparation conditions, we aim to identify and address unintentional variations that may arise during the preparation process. Our findings highlight the importance of carefully designing experiments to avoid unintentional variation, thus contributing to a rigorous investigation for a comprehensive understanding of phosphors.

2. Results and Discussion
2.1. Grinding

We first discuss tolerance, defined as the permissible error, for analyzing the ratio of CTB to f-f transition. Figure 1a shows the photoluminescence excitation (PLE) spectra of two CYASO:Eu samples prepared individually (by two different individuals). The PLE spectra are normalized to the highest peak intensity in each spectrum (here, the intensity of the f-f transition at 395 nm) to exclude the error due to mass variation between measurements. Both samples exhibit different CTB intensities, consistent across repeated measurements (standard deviation ≈ 0.01). Typically, in sample preparation using a solid-state reaction, the grinding process is performed after heating to eliminate lumps and ensure homogeneity. We initially hypothesized that differences in grinding conditions (e.g., force, duration, concentration) by different individuals might account for the observed discrepancy. To check this, we further performed grinding processes for 10 min and 20 min for the same individual. The ratio slightly increases continuously as the grinding process is executed further, likely due to the reduction in oxygen vacancies on the particle surface during the grinding process [19,20]. Although the ratios slightly increase, the gap between the ratios of the two different samples almost persists (Figure 1b). This result suggests that unintentional variation can arise at any stage during sample preparation using a solid-state reaction. The difference observed in the two different samples is considered as the tolerance (~0.1) in analyzing the ratio of CTB to f-f transition, provided the grinding processes are consistently executed. Furthermore, we are not able to see a significant difference in the Judd–Ofelt parameters and asymmetric ratios (i.e., R-factor) across the samples (Table 1), indicating that the local environment of Eu ions is insensitive to the grinding process (see Figure S1 and Table S1 in Supplementary Information for details). The ratio of CTB to f-f transition seems much more sensitive to the sample preparation than optical transition strength parameters. Hereinafter, we investigate the possible sources causing the change in PL(E) spectra while considering the tolerance discussed here.

Table 1. Optical transition strength parameters and asymmetric ratios (R-factors) of two CYASO:Eu samples prepared under the same conditions nominally.

Samples	Grinding Time (min)	Ω_2 (10^{-20} cm^2)	Ω_4 (10^{-20} cm^2)	R-Factor
As prepared #1	5	3.484	2.467	2.284
	10	3.448	2.367	2.260
	20	3.435	2.374	2.251
As prepared #2	5	3.498	2.466	2.294
	10	3.374	2.269	2.212
	20	3.428	2.352	2.249

subjected to different milling times, the samples would be prepared with different Al contents. Figure 3b shows the XRD patterns of the CYASO:Eu prepared using the Al_2O_3 powders subjected to different milling times. No significant difference is observed for the CYASO phase, but the peak intensity of the secondary phase, Y_2O_3, slightly increases as the ball milling time increases. The formation of secondary phases might seem correlated with the particle size of Al_2O_3 or simply ball milling time. The PL(E) spectra of the CYASO:Eu samples are represented in Figure 4a. The overall intensity of both PLE and PL spectra decreases, and the ratio of CTB to f-f transition increases as the ball milling time increases (Figure 4b), while no meaningful differences are observed for the optical transition strength parameters and R-factors within the tolerance (Table 2). The photoluminescent properties might also seem correlated with the ball milling time. *This is wrong*—the observed changes in both structural and optical measurements are due to off-stoichiometry. This will be evident in the next section. Many sources cause contamination in the ball milling process, and they can lead to misleading or incorrect interpretations.

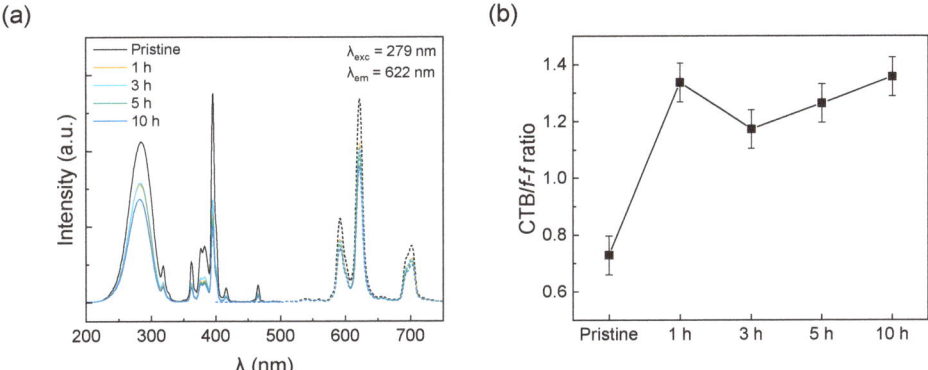

Figure 4. (**a**) Photoluminescence (excitation) spectra of the Si-substituted $CaYAlO_4$:Eu (CYASO:Eu) samples prepared with the Al_2O_3 powders subjected to the different ball milling times (0 to 10 h). The photoluminescence excitation (PLE) and PL spectra were monitored at 622 nm emission (λ_{em}, solid lines) and 279 nm excitation (λ_{exc}, dashed lines), respectively. (**b**) The ratios of charge transfer band (CTB) to f-f transition of the CYASO:Eu samples as the function of ball milling time for Al_2O_3 powders.

Table 2. Optical transition strength parameters and asymmetric ratios (R-factors) of the Si-substituted $CaYAlO_4$:Eu (CYASO:Eu) samples prepared with the Al_2O_3 powders subjected to the different ball milling times.

Ball Milling Time	Ω_2 (10^{-20} cm^2)	Ω_4 (10^{-20} cm^2)	R-Factor
10 h	3.474	2.531	2.295
5 h	3.483	2.550	2.300
3 h	3.493	2.614	2.307
1 h	3.463	2.531	2.287
Pristine	3.498	2.466	2.294

2.3. Precursors

To see the effect of different Al precursors on the photoluminescent properties of CYASO:Eu while avoiding potential off-stoichiometry issues, we used entire powders after a specific procedure to manipulate the precursors. We initially determined the masses of the starting materials considering the stoichiometry and performed the ball milling process on Al_2O_3 powder. In the ball milling, the overall mass of the powder increases because of the organic scraps from the container, while the mole number of Al remains consistent. To avoid any loss of the ball-milled powders, we did not collect the powders from the container, and we put the other starting materials into the container and performed the mixing procedure. We applied a similar strategy for preparing Al precursor from the solution of $Al(NO_3)_3 \cdot 9H_2O$ (aluminum nitrate nonahydrate); the Al precursor obtained by this method was expected to have very small grain and fine particle size. The mass of $Al(NO_3)_3 \cdot 9H_2O$ powder was first determined considering the stoichiometry, and the powder was dissolved in distilled water. The solution was poured into a milling jar and dried using an oven. The Al precursor prepared by drying the solution was mixed with the other starting materials. The XRD patterns of the Al precursors used in this study are shown in Figure 5a. The pristine $Al(NO_3)_3 \cdot 9H_2O$ powder was also used as one of the Al precursors for comparison. It can be seen that the ball-milled Al_2O_3 powder exhibits broader XRD peak widths, indicating reduced particle size, and the powder obtained by drying the $Al(NO_3)_3 \cdot 9H_2O$ solution does not show any diffraction peaks, as expected for amorphous or very small grain powders.

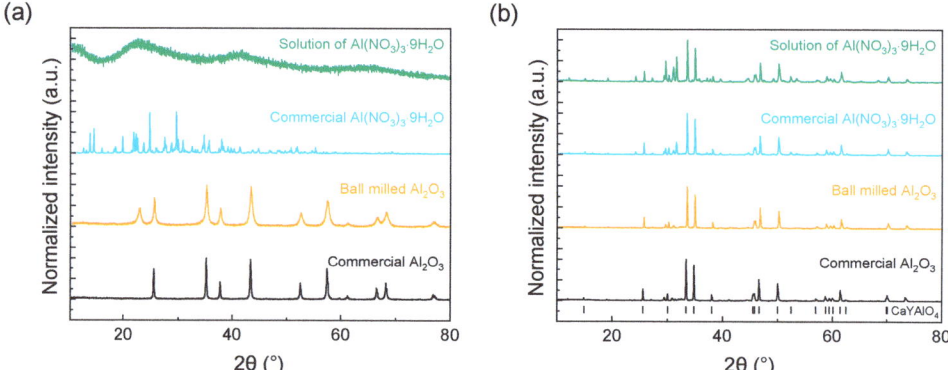

Figure 5. (**a**) X-ray diffraction (XRD) patterns of Al precursors: commercial Al_2O_3, ball-milled Al_2O_3, commercial $Al(NO_3)_3 \cdot 9H_2O$, and powder obtained by drying the solution of $Al(NO_3)_3 \cdot 9H_2O$. (**b**) XRD patterns of the Si-substituted $CaYAlO_4$:Eu (CYASO:Eu) samples prepared with the different Al precursors. The black bars indicate the reference peak positions of $CaYAlO_4$ (JCPDS No. 81-0742).

The XRD patterns of the CYASO:Eu prepared with the different Al precursors are presented in Figure 5b. First, we note the very similar diffraction patterns regardless of the ball milling of Al_2O_3 powder. This result attests that the previous results observed in Figures 3b and 4 are indeed due to off-stoichiometry. Notably, the samples prepared with $Al(NO_3)_3 \cdot 9H_2O$ show the presence of an additional secondary phase, $Ca_2Al_2SiO_7$, compared to the sample prepared with the Al_2O_3 powders. No significant difference is observed for the CYAO phase.

The PL(E) spectra of CYASO:Eu prepared with the different Al precursors are presented in Figure 6a. The overall intensities seem dependent on the precursors, but the quantum efficiencies of the samples were found to be similar; we thus attribute the difference to the mass difference used for each measurement. The difference in the ratio of CTB to *f-f* transition is insignificant considering the tolerance as discussed above (Figure 6b). The optical transition strength parameters and *R*-factors of the CYASO:Eu samples are listed in Table 3. Except for the sample prepared with the $Al(NO_3)_3 \cdot 9H_2O$ solution, all samples exhibit very similar optical strength parameters and *R*-factors. Ω_2 parameter depends on the local crystal environment of rare earth ion sites; Ω_4 and Ω_6 are related to the viscosity and rigidity of a host matrix. The CYASO:Eu sample prepared with the $Al(NO_3)_3 \cdot 9H_2O$ solution exhibits the considerable formation of secondary phases compared to the other samples. The formation of secondary phases may affect both stoichiometry and defect levels of a desired compound, and/or directly host the luminescent activator with a different local structure and environment. Further study is required to elucidate the effect of precursors on the luminescent properties. This study warns of possible misleading due to unintentional variation atop intended systematic variation. We need to pay attention when designing sample preparation protocols involving systematic variation to minimize unintentional variation across individual sample preparations. Combinatorial sample preparation with the slogan "many at a time" serves as a valuable approach not only for conducting high-throughput experiments but also for facilitating systematic studies while excluding unintended variation [22–26].

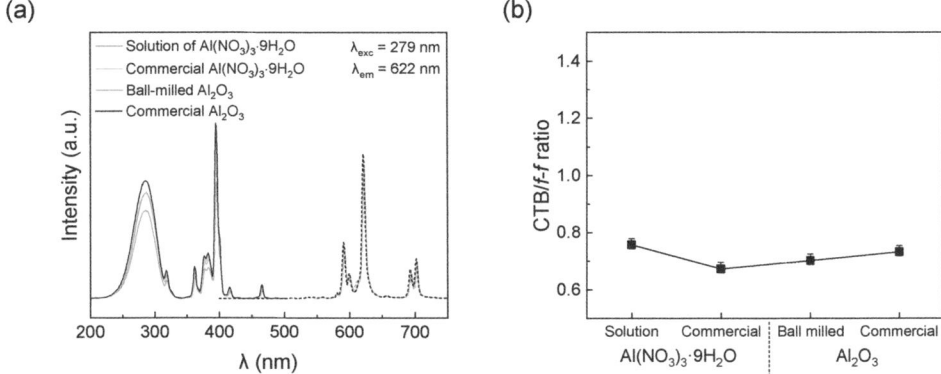

Figure 6. (a) Photoluminescence (excitation) spectra of the Si-substituted CaYAlO$_4$:Eu (CYASO:Eu) samples prepared with the different Al precursors (see the main text). The photoluminescence excitation (PLE) and PL spectra were monitored at 622 nm emission (λ_{em}, solid lines) and 279 nm excitation (λ_{exc}, dashed lines), respectively. (b) The ratios of charge transfer band (CTB) to f-f transition of the CYASO:Eu samples.

Table 3. Optical transition strength parameters and asymmetric ratios (R-factors) of the Si-substituted CaYAlO$_4$:Eu (CYASO:Eu) samples prepared with the different Al precursors.

Precursors		Ω_2 (10^{-20} cm^2)	Ω_4 (10^{-20} cm^2)	R-Factor
Al$_2$O$_3$	Commercial	3.498	2.466	2.294
	Ball-milled	3.502	2.415	2.298
Al(NO$_3$)$_3$·9H$_2$O	Commercial	3.472	2.414	2.282
	Solution	3.355	2.121	2.225

3. Experimental Methods

Sample preparation: Samples with a composition of Ca$_{1.15}$Y$_{0.8}$Al$_{0.8}$Si$_{0.2}$O$_4$:Eu$_{0.05}$ (CYAGO:Eu) were prepared through a solid-state reaction method using CaCO$_3$ (99.95%, Alfa Aesar, USA), Y$_2$O$_3$ (99.99%, Sigma Aldrich, St. Louis, MO, USA), SiO$_2$ (99.998%, Acros Organics, Waltham, MA, USA), and Eu$_2$O$_3$ (99.99%, Alfa Aesar, Ward Hill, MA, USA); as Al precursors, Al$_2$O$_3$ (99.95%, Sigma Aldrich, St. Louis, MO, USA) and Al(NO$_3$)$_3$·9H$_2$O (99.99%, Sigma Aldrich, St. Louis, MO, USA) were used (see the main text for details). The mixtures of the starting materials were subjected to a planetary ball milling process using ZrO$_2$ or Al$_2$O$_3$ balls for 3 h and subsequently heated at 1400 °C using a box furnace. After the synthesis process, the obtained powders were ground in an agate mortar with a pestle for 10 min to remove lumps and ensure homogeneity.

Characterizations: X-ray diffraction (XRD) measurements were carried out using an Ultima IV (Rigaku, Tokyo, Japan) with Cu Kα_1 = 1.5406 Å. The diffraction patterns were collected in the 2θ range of 10–80° with a step size of 0.02°. Photoluminescence (PL) and PL excitation (PLE) measurements were performed using a Photon Technology International (PTI) spectrophotometer equipped with a 60 W Xe-arc lamp.

4. Conclusions

When preparing phosphor samples using a solid-state reaction, maintaining consistency across multiple sample preparations for systematic studies can be challenging due to the numerous steps involved. Inconsistent grinding processes can lead to unintentional changes in photoluminescence spectra, likely due to the oxidation of particle surfaces. Additionally, the possible contamination from ball and container scraps during the ball milling process may result in mass change and significant off-stoichiometry issues if the process is carried out for precursors individually. These issues can be addressed by implementing carefully designed sample preparation protocols, which enable systematic investigation, like the present study on the effect of different Al precursors. This study highlights the potential for unintentional variation to obscure systematic variation and emphasizes the importance of meticulous sample preparation, awareness of potential sources of error, and the determination of statistical error bars in systematic studies.

Supplementary Materials: The following supporting information can be downloaded at: https://www.mdpi.com/article/10.3390/inorganics12060150/s1, Figure S1: Photoluminescence spectra of a Si-substituted CaYAlO$_4$:Eu (CYASO) sample.; Table S1: Optical transition strength parameters and asymmetric ratios (R-factors) of a Si-substituted CaYAlO$_4$:Eu (CYASO:Eu) sample.

Author Contributions: Conceptualization, J.H.O. and S.L.; validation, J.H.O.; investigation, J.H.O., Y.L., J.K and W.T.H.; resources, H.K.Y.; writing—original draft preparation, J.H.O. and S.L.; writing—review and editing, M.K. and S.L.; visualization, J.H.O.; supervision, S.L.; funding acquisition, J.H.O., M.K. and S.L. All authors have read and agreed to the published version of the manuscript.

Funding: This work was supported by the National Research Foundation of Korea Grant funded by the Korean Government (NRF-2021R1C1C1009863). M. Kang acknowledges support from National Research Foundation of Korea Grant funded by the Korean Government (RS-2024-00358042). J. H. Oh acknowledges support from National Research Foundation of Korea Grant funded by the Korean Government (RS-2023-00248068).

Data Availability Statement: The data presented in this study are available on request from the corresponding author.

Conflicts of Interest: The authors declare no conflict of interest.

References

1. Nair, G.B.; Swart, H.C.; Dhoble, S.J. A review on the advancements in phosphor-converted light emitting diodes (pc-LEDs): Phosphor synthesis, device fabrication and characterization. *Prog. Mater. Sci.* **2020**, *109*, 100622. [CrossRef]
2. Peng, Y.; Wang, H.; Liu, J.; Sun, Q.; Mou, Y.; Guo, X. Broad-Band and Stable Phosphor-in-Glass Enabling Ultrahigh Color Rendering for All-Inorganic High-Power WLEDs. *ACS Appl. Electron. Mater.* **2020**, *2*, 2929–2936. [CrossRef]
3. Wu, Z.; Li, C.; Zhang, F.; Huang, S.; Wang, F.; Wang, X.; Jiao, H. High-performance ultra-narrow-band green-emitting phosphor LaMgAl$_{11}$O$_{19}$:Mn^{2+} for wide color-gamut WLED backlight displays. *J. Mater. Chem. C* **2022**, *10*, 7443–7448. [CrossRef]
4. Li, J.; Yan, J.; Wen, D.; Khan, W.U.; Shi, J.; Wu, M.; Su, Q.; Tanner, P.A. Advanced red phosphors for white light-emitting diodes. *J. Mater. Chem. C* **2016**, *4*, 8611–8623. [CrossRef]
5. Xu, Y.; Zhang, L.; Yin, S.; Wu, X.; You, H. Highly efficient green-emitting phosphors with high color rendering for WLEDs. *J. Alloys Compd.* **2022**, *911*, 165149. [CrossRef]
6. Li, Y.; Yin, Y.; Wang, T.; Wu, J.; Zhang, J.; Yu, S.; Zhang, M.; Zhao, L.; Wang, W. Ultra-bright green-emitting phosphors with an internal quantum efficiency of over 90% for high-quality WLEDs. *Dalton Trans.* **2021**, *50*, 4159–4166. [CrossRef] [PubMed]
7. Wang, Z.; Yang, Z.; Wang, N.; Zhou, Q.; Zhou, J.; Ma, L.; Wang, X.; Xu, Y.; Brik, M.G.; Dramićanin, M.D.; et al. Single-Crystal Red Phosphors: Enhanced Optical Efficiency and Improved Chemical Stability for wLEDs. *Adv. Opt. Mater.* **2020**, *8*, 1901512. [CrossRef]
8. Leng, Z.; Bai, H.; Qing, Q.; He, H.; Hou, J.; Li, B.; Tang, Z.; Song, F.; Wu, H. A Zero-Thermal-Quenching Blue Phosphor for Sustainable and Human-Centric WLED Lighting. *ACS Sustain. Chem. Eng.* **2022**, *10*, 10966–10977. [CrossRef]
9. Li, J.; Liang, Q.; Hong, J.-Y.; Yan, J.; Dolgov, L.; Meng, Y.; Xu, Y.; Shi, J.; Wu, M. White Light Emission and Enhanced Color Stability in a Single-Component Host. *ACS Appl. Mater. Interfaces* **2018**, *10*, 18066–18072. [CrossRef]

10. Dai, P.; Wang, Q.; Xiang, M.; Chen, T.-M.; Zhang, X.; Chiang, Y.-W.; Chan, T.-S.; Wang, X. Composition-driven anionic disorder-order transformations triggered single-Eu^{2+}-converted high-color-rendering white-light phosphors. *Chem. Eng. J.* **2020**, *380*, 122508. [CrossRef]
11. Tyagi, A.; Nigam, S.; Sudarsan, V.; Majumder, C.; Vatsa, R.K.; Tyagi, A.K. Why Do Relative Intensities of Charge Transfer and Intra-4f Transitions of Eu^{3+} Ion Invert in Yttrium Germanate Hosts? Unravelling the Underlying Intricacies from Experimental and Theoretical Investigations. *Inorg. Chem.* **2020**, *59*, 12659–12671. [CrossRef] [PubMed]
12. Otsuka, T.; Oka, R.; Hayakawa, T. Eu^{3+} Site Distribution and Local Distortion of Photoluminescent Ca_3WO_6:(Eu^{3+}, K^+) Double Perovskites as High-Color-Purity Red Phosphors. *Adv. Sci.* **2023**, *10*, 2302559. [CrossRef] [PubMed]
13. Judd, B.R. Optical absorption intensities of rare-earth ions. *Phys. Rev.* **1962**, *127*, 750–761. [CrossRef]
14. Ofelt, G.S. Intensities of Crystal Spectra of Rare-Earth Ions. *J. Chem. Phys.* **1962**, *37*, 511–520. [CrossRef]
15. Sreena, T.S.; Raj, A.K.V.; Rao, P.P. Effects of charge transfer band position and intensity on the photoluminescence properties of $Ca_{1.9}M_2O_7$:$0.1Eu^{3+}$ (M = Nb, Sb and Ta). *Solid State Sci.* **2022**, *123*, 106783. [CrossRef]
16. Gupta, S.K.; Gupta, R.; Vats, B.G.; Gamare, J.S.; Kadam, R.M. Inversion in usual excitation intensities from solid state phosphor and improved fluorescence of Eu^{3+} ion in type (IV) deep eutectic solvent. *J. Lumin.* **2021**, *235*, 118026. [CrossRef]
17. Sreevalsa, S.; Ranjith, P.; Ahmad, S.; Sahoo, S.K.; Som, S.; Pandey, M.K.; Das, S. Host sensitized photoluminescence in $Sr_{2.9-3x/2}Ln_xAlO_4F$: $0.1\ Eu^{3+}$ (Ln = Gd, Y) for innovative flexible lighting applications. *Ceram. Int.* **2020**, *46*, 21448–21460. [CrossRef]
18. Zhao, Q.; Qian, B.; Wang, Y.; Duan, T.; Zou, H.; Song, Y.; Sheng, Y. Facile synthesis of CaO:Eu^{3+} and comparative study on the luminescence properties of CaO:Eu^{3+} and $CaCO_3$:Eu^{3+}. *J. Lumin.* **2022**, *241*, 118491. [CrossRef]
19. Zhao, S.; Peng, Y. The oxidation of copper sulfide minerals during grinding and their interactions with clay particles. *Powder Technol.* **2012**, *230*, 112–117. [CrossRef]
20. Gao, Y.; Zhu, X.; Shi, H.; Jiang, P.; Cong, R.; Yang, T. Eu^{3+} and Tb^{3+} doped $LiCaY_5(BO_3)_6$: Efficient red and green phosphors under UV or NUV excitations. *J. Lumin.* **2022**, *242*, 118598. [CrossRef]
21. Sari, A.; Keddam, M.; Guittoum, A. Effect of iron impurity on structural development in ball-milled ZrO_2–3mol% Y_2O_3. *Ceram. Int.* **2015**, *41*, 1121–1128. [CrossRef]
22. Li, M.-X.; Zhao, S.-F.; Lu, Z.; Hirata, A.; Wen, P.; Bai, H.-Y.; Chen, M.; Schroers, J.; Liu, Y.; Wang, W.-H. High-temperature bulk metallic glasses developed by combinatorial methods. *Nature* **2019**, *569*, 99–103. [CrossRef] [PubMed]
23. Yuan, J.; Chen, Q.; Jiang, K.; Feng, Z.; Lin, Z.; Yu, H.; He, G.; Zhang, J.; Jiang, X.; Zhang, X.; et al. Scaling of the strange-metal scattering in unconventional superconductors. *Nature* **2022**, *602*, 431–436. [CrossRef] [PubMed]
24. Liang, Y.G.; Lee, S.; Yu, H.S.; Zhang, H.R.; Liang, Y.J.; Zavalij, P.Y.; Chen, X.; James, R.D.; Bendersky, L.A.; Davydov, A.V.; et al. Tuning the hysteresis of a metal-insulator transition via lattice compatibility. *Nat. Commun.* **2020**, *11*, 3539. [CrossRef] [PubMed]
25. Nam, K.; Oh, J.H.; Bae, J.-S.; Lee, S. Effects of Heat Treatment on the Microstructure and Optical Properties of Sputtered GeO_2 Thin Films. *Adv. Eng. Mater.* **2023**, *25*, 2300456. [CrossRef]
26. Kim, H.; Nam, K.; Park, J.; Kang, M.; Bae, J.-S.; Hong, W.T.; Yang, H.K.; Jeong, J.H.; Oh, J.H.; Lee, S. Hydrogen-mediated manipulation of luminescence color in single-component Eu doped $CaYAlSiO_4$ by defect passivation. *J. Alloys Compd.* **2023**, *932*, 167610. [CrossRef]

Disclaimer/Publisher's Note: The statements, opinions and data contained in all publications are solely those of the individual author(s) and contributor(s) and not of MDPI and/or the editor(s). MDPI and/or the editor(s) disclaim responsibility for any injury to people or property resulting from any ideas, methods, instructions or products referred to in the content.

The Effect of Organic Spacer Cations with Different Chain Lengths on Quasi-Two-Dimensional Perovskite Properties

Lei Zhang [1], Mingze Xia [2], Yuan Zhang [1], Li Song [3], Xiwei Guo [3], Yong Zhang [1], Yulei Wang [1] and Yuanqin Xia [1,*]

[1] Hebei Key Laboratory of Advanced Laser Technology and Equipment, School of Electronics and Information Engineering, Hebei University of Technology, 5340 Xiping Road, Tianjin 300401, China; zl98nian@163.com (L.Z.); zhangyuan956912516@163.com (Y.Z.); zhangyong@hebut.edu.cn (Y.Z.); wyl@hebut.edu.cn (Y.W.)
[2] Cnooc Energy Technology and Services, Tianjin 300450, China; xiamz2023@163.com
[3] Tianjin Key Laboratory of Electronic Materials and Devices, School of Electronics and Information Engineering, Hebei University of Technology, 5340 Xiping Road, Tianjin 300401, China; songli@hebut.edu.cn (L.S.); g17634324378@163.com (X.G.)
* Correspondence: xiayq@hebut.edu.cn

Abstract: In the past 20 years, perovskite-related research has attracted wide attention. The related research into two-dimensional/quasi-two-dimensional perovskite has propelled the research of perovskite materials to a new height. To improve the properties of quasi-2D perovskite, improve the stability of materials, and achieve specific functions, using different types, volumes, and lengths of organic spacers is an essential method. In this paper, quasi-2D perovskites with EDA (ethylene diammonium), PDA (1,3-propanediammonium), and BDA (1,4-butanediammonium) (m = 2–4) as organic spacers were prepared, and the effects of different organic spacers on the 2D perovskite were investigated. The results show that the length of the organic spacer significantly impacts the perovskite's properties. A shorter organic spacer can effectively reduce the quantum confinement and dielectric confinement in perovskite. It should be noted that if the organic spacer is too short, the stability of the quasi-2D perovskite will be greatly reduced.

Keywords: quasi-2D perovskite; transient absorption spectra; carrier dynamics

1. Introduction

With the development of human society, the living standard of human beings is constantly improving, accompanied by the rapid consumption of energy. The utilization of renewable energy has become an important way to solve this problem. Perovskite is starting to attract attention in this context. Because of its excellent performance and low cost, it has become the research hotspot of many scholars. It has shown a broad application prospect in many fields, such as solar cells [1–10], LEDs [11–21], and so on. In order to make perovskite materials achieve specific functions and improve their efficiency, it is very important to understand the carrier dynamics of perovskite materials, which have an important guiding role.

In order to obtain high-stability and high-efficiency perovskite materials, research on 2D/quasi-2D perovskite has been an important direction. The basic properties of Dion–Jacobson (DJ) perovskite are usually determined by organic spacer cations [22–24]. Among these organic spacer cations, diammonium is the most widely studied, including linear cations [25,26] and ring cations [27–29]. The ammonium cation of an organic spacer cation can form hydrogen bonds with multiple ends of the inorganic octahedral structure, so factors such as the spatial configuration and length of the organic spacer cation are crucial to the structure and size of perovskite [30–32]. It also directly affects the distortion of the inorganic layer connected by the organic spacer [33,34]. In contrast to three-dimensional perovskite, quasi-2D perovskite has quantum confinement and dielectric confinement [35–37]. The quantum confinement and dielectric confinement lead to a degradation of the properties of quasi-2D perovskite. There are two ways to solve this problem: using a shorter and higher-dielectric-constant organic isolation layer [38] or improving the cross-layer charge transfer [39]. It has been found that large organic alkyl ammonium cations can achieve a uniform distribution of quantum wells (QWs) in perovskite and create a flat energy landscape [40]. Kanatzidis studied DJ-type perovskites with organic spacers of different chain lengths (m = 4–9) and proved that the carbon chain length of alkyl diammonium cations can affect the optical properties of perovskite materials by adjusting the spacing between inorganic layers [41]. Using smaller organic spacer cations instead of larger organic spacer cations in order to shorten the distance between inorganic layers can effectively weaken the quantum confinement effect in perovskite [42]. In other aspects, Ahmacl et al. [43] and Zheng et al. [44] used organic isolation layers to adjust and control the crystallinity, charge mobility, QW width, and distribution of aligned 2D perovskites. However, researchers still lack understanding regarding the effect of organic spacers of different lengths on quasi-2D perovskites. The internal mechanisms of quasi-2D perovskite properties (such as the energy-transfer processes) induced by different organic spacers still need to be further explored.

In order to further explore the effects of different organic spacer cations on the carrier dynamics of quasi-2D perovskite, we prepared quasi-2D perovskite thin films with $CsPbBr_3$ as a matrix and EDA^{2+} (ethylene diammonium), PDA^{2+} (1,3-propanediammonium), and BDA^{2+} (1,4-butanediammonium) as organic spacer cations. As shown in Figure 1, perovskites with different n values also have some differences in structure. The quasi-2D perovskite film used in this paper is made using the spin-coating method, and the n value is four.

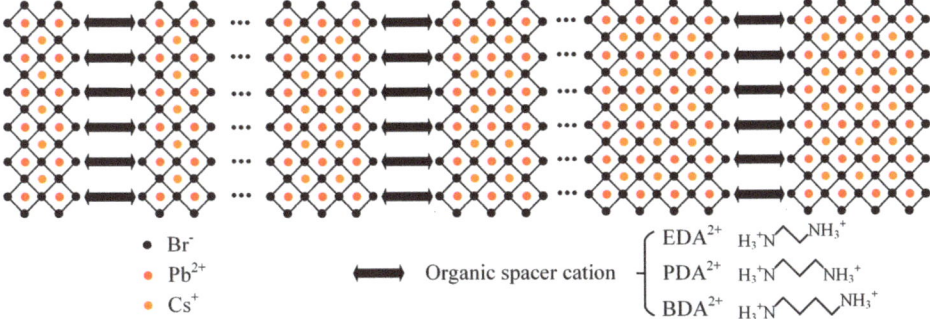

Figure 1. $EDACs_3Pb_4Br_{13}$, $PDACs_3Pb_4Br_{13}$, and $BDACs_3Pb_4Br_{13}$ diagram of multi-n phase structure.

2. Results

In this study, the films of $EDACs_3Pb_4Br_{13}$, $PDACs_3Pb_4Br_{13}$, and $BDACs_3Pb_4Br_{13}$ are denoted as EDA, PDA, and BDA, respectively. The composition and content of quasi-2D perovskite were examined using ultraviolet–visible spectroscopy (UV–Vis absorption spectrum).

Figure 2a illustrates the exciton absorption signals of EDA, PDA, and BDA. Notably, a prominent exciton absorption peak is observed near 480 nm in all three samples. This specific exciton absorption peak signal aligns with the absorption characteristics of the four-dimensional phase (n = 4) [40].

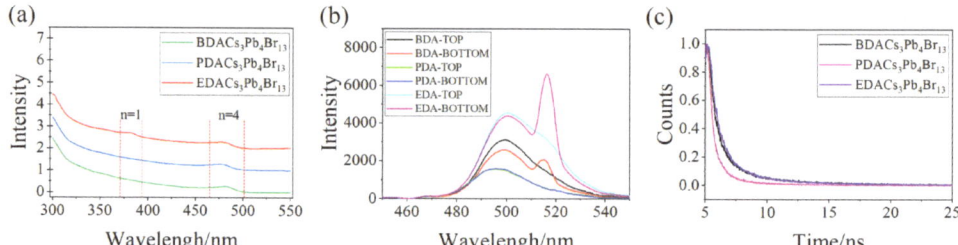

Figure 2. EDACs$_3$Pb$_4$Br$_{13}$, PDACs$_3$Pb$_4$Br$_{13}$, and BDACs$_3$Pb$_4$Br$_{13}$ (**a**) UV−Vis absorption spectra. (**b**) PL spectra. (**c**) TRPL spectra.

In perovskite materials, the notation $n = 4$ signifies the presence of the $n = 4$ phase within the perovskite film, indicating that the film consists of four inorganic layers. This designation holds for any other value of n as well. Notably, the exciton absorption peaks at 480 nm for the $n = 4$ phase in different organic spacers are nearly identical, suggesting their four-dimensional phase band gaps are closely aligned. Moreover, the EDA film sample exhibits a notable exciton absorption peak at 381 nm, corresponding to the one-dimensional phase absorption of the EDA film. In forming a DJ-type perovskite, the precursor solution is tailored according to the specific n value desired for the perovskite. It is worth noting that other dimensional phase components may also emerge in the final production of perovskite. While the UV−Vis spectra do not show exciton absorption signals of other dimensional phases, this does not imply the absence of low-dimensional phase components in quasi-2D perovskite films. These findings suggest that the enhanced interlayer interaction in EDA, PDA, and BDA films mitigates the quantum confinement effect in the quasi-2D phase. Furthermore, short-chain organic spacers in EDA effectively reduce the distortion of DJ-type perovskite [45–47]. The quantum confinement effect (QCE) is a phenomenon in which the quantization of energy in microscopic particles becomes more pronounced as the spatial confinement of their motion decreases. In the context of electrons in a solid, when at least one dimension in the three-dimensional scale reaches the nanometer scale, the movement of electrons in that dimension becomes restricted, leading to the manifestation of quantum effects. This confinement results in an increase in the energy-level band gap. In the case of perovskite films, this effect leads to a uniform distribution of QWs. Comparing RP perovskite with DJ perovskite, it is noted that DJ perovskite has fewer organic spacers. This reduction in organic spacers is suggested as one of the reasons why the exciton absorption peak of DJ perovskite is not as prominent.

Additionally, subtle differences exist despite the close proximity of the $n = 4$ exciton absorption peaks in quasi-2D organic spacer cations of perovskite films. Specifically, the $n = 4$ phase exciton absorption peak of PDA exhibits a 4 nm blue shift compared to the other two samples. Examining the exciton absorption of EDA at 380 nm in UV−Vis spectra, it is observed that EDA demonstrates higher exciton absorption at 380 nm. This suggests that EDA^{2+}, compared to PDA^{2+} and BDA^{2+}, can promote the formation of a low-dimensional phase with small n values.

The exciton absorption peaks of perovskite films are initially observable through UV−Vis spectroscopy, providing valuable information on the absorption characteristics of the sample. In Figure 2b, the photoluminescence (PL) spectra of quasi-2D perovskite films with different organic spacer cations are presented. The three samples are excited from the top, and the PL spectra collected from the bottom of the perovskite film are analyzed. The PL spectrum signal exhibits two main parts. First, there is a broad peak near 500 nm in the left part of the spectrum. Combined with UV−Vis spectra, it can be deduced that this broad peak corresponds to the fluorescence signal of the $n = 4$ phase. The second part is a narrow peak near 525 nm in the right part of the spectrum. This peak is attributed to the amplified spontaneous emission (ASE) signal of the quasi-2D perovskite samples. The ASE signal in the PL spectrum is produced by the bulk phase formed in quasi-2D perovskite. In the excited state, carriers spontaneously radiate to the ground state, releasing photons. The energy of these photons corresponds to the energy difference between the excited-state carrier and the ground-state carrier. The photons produced by spontaneous emission can induce stimulated emission of carriers in the bulk phase, and the effect is that the spontaneous emission signal is amplified, and, thus, the ASE signal is produced. Although the ASE signal contains stimulated radiation, it is still spontaneous radiation in nature. The amplification of spontaneous emission is directly related to the number of particles in the upper energy level. From the PL spectra, EDA perovskite film samples are more likely to produce an ASE signal than BDA and PDA samples. This is because the energy transfer in EDA film samples is more intense and smoother. The carrier in the $n = 4$ phase can transfer into the bulk phase rapidly and effectively. The carrier number can be rapidly accumulated in the bulk phase. Compared to the PL spectrum without an ASE signal, the decrease in the fluorescence signal of the $n = 4$ phase when the ASE signal is generated also proves this point. From the PL spectra, it can be seen that the ASE signal is more easily generated by the excitation of the quasi-2D perovskite samples from the bottom. The energy-transfer process from the low-dimensional phase to the bulk phase at the bottom of perovskite films is better than that at the top of the perovskite film. We attribute this phenomenon to the higher density of states at the bottom of perovskite films. This is due to the deposition of partial phases on the bottom of the films under gravity during spin coating. The next experiments will also prove it.

In Figure 2b, the PL spectra of the perovskite films reveal a notable blue shift of 4 nm in the PL emission peak of PDA perovskite films compared to BDA and EDA. This shift aligns with the findings from the UV−Vis spectra. The origin of this phenomenon is the distortion of the perovskite lattice induced by PDA^{2+} organic spacer cations. The distortion alters the Pb−Br bond length in quasi-2D perovskite, resulting in a modification of the perovskite's band gap to some extent. Referencing the literature [48], when the organic spacer cations are PDA^{2+}, the distortion in the perovskite is more pronounced compared to with BDA^{2+} and EDA^{2+} organic spacer cations. For PDA, this substantial lattice distortion is responsible for the observed blue shift in the PL spectral emission peak. Furthermore, this lattice distortion is also responsible for the weak $n = 4$ phase fluorescence signal of PDA. When the organic spacer cations are BDA^{2+} and EDA^{2+}, the perovskite lattice also undergoes distortion, albeit less severely than with PDA^{2+}. Consequently, the band gap of the $n = 4$ phase experiences minimal change. Notably, EDA^{2+} cations, being shorter in length compared to BDA^{2+}, would ideally result in a slightly narrower $n = 4$ phase gap. However, the lattice distortion caused by EDA^{2+} is slightly greater than that induced by BDA^{2+}. These two factors compensate for each other, resulting in a nearly identical PL peak. From Figure 2b, it is clear that the PL peak of EDA is significantly higher than BDA. This is because the shorter chain length of EDA^{2+} reduces the distance between inorganic layers in quasi-2D perovskite and weakens the quantum confinement effect and dielectric confinement in quasi-2D perovskite. This results in lower exciton binding energy, which is beneficial for improving the carrier separation rate in quasi-2D perovskite [49].

As shown in the Time-Resolved Photoluminescence (TRPL) spectra of perovskite films in Figure 2c, the decay of PDA perovskite films is the most rapid, and the attenuation curves of BDA and EDA do not have many differences. The fluorescence lifetimes of the BDA, PDA, and EDA films were 3.6 ns, 1.9 ns, and 3.5 ns, respectively. The fluorescence lifetime of PDA can also be attributed to the decrease in radiation recombination due to the lattice distortion caused by organic spacer cations. The lattice distortion caused by BDA^{2+} is slightly less than that caused by EDA^{2+}. The difference in fluorescence lifetime between BDA and EDA also verifies the above conclusion.

In order to analyze the effect of organic spacers of different lengths on quasi-2D perovskite further, we collected the transient absorption spectra of three perovskite film samples. As shown in Figure 3, the transient absorption spectrum mainly consists of two parts; the positive signal of the transient absorption spectrum is a broad peak covering a range of tens of nanometers, and the negative signal in the transient absorption spectrum is the ground-state bleaching signal (GSB) of the $n = 4$ phase. Although the broad peaks are in the form of positive signals, they also contain GSB signals of two-dimensional and three-dimensional phases. The intensity of the excited-state absorption (ESA) signal is stronger than the intensity of the GSB signal, so the ESA signal covers the negative signal of the ground-state bleaching signal, and the transient absorption spectrum shows as a positive signal. In addition, the Stark effect is also one of the reasons why the GSB signal of the two-dimensional and three-dimensional phases is covered. The positive signal mainly appears around the GSB signal of the $n = 4$ phase, which is caused by the absorption change of material under the influence of the electric field [50–52]. It is obvious that the transient absorption spectra under positive excitation are much weaker than those under negative excitation. In combination with PL spectra and the above analysis, the reason for the above phenomenon is that the density of states at the bottom of the perovskite film is higher than that at the top of the film. In the process of preparing perovskite films by spin coating the perovskite precursor solution, the perovskite films began to crystallize at the liquid−gas interface. However, due to the influence of gravity, the quasi-2D phase, which should be uniformly distributed, is inevitably deposited, and the density of states at the bottom of the perovskite film increases.

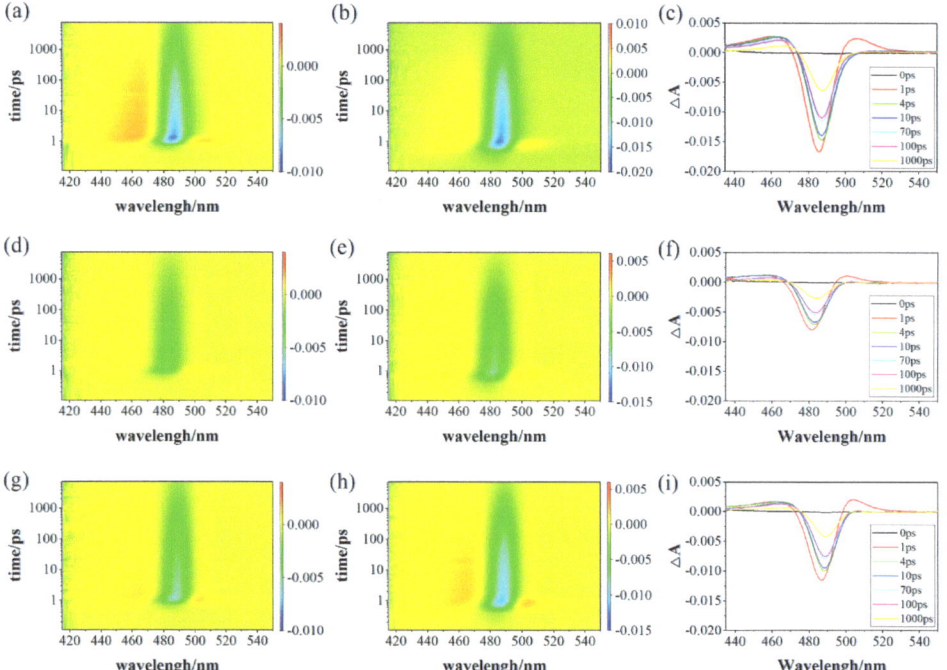

Figure 3. (a,b) are transient absorption spectra of EDA pumped front and back. (d,e) are transient absorption spectra of PDA pumped front and back. (g,h) are transient absorption spectra of BDA pumped front and back. (c,f,i) are transient absorption spectra of EDA, PDA, and BDA pumped back with different time delays.

The transient absorption spectra under positive excitation for perovskite films exhibit similarities to those under negative excitation. We extracted the transient absorption spectra for both top-excited and bottom-excited films, and the results are presented in Figure 4a and 4b, respectively. Moreover, the relationship between the transient absorption spectra of different organic spacer cationic perovskite materials excited from the top and bottom of the film is also similar. The time constants of monomolecular recombination (defect recombination), bimolecular recombination (radiation recombination), and auger recombination can be obtained by three-exponential fitting of the transient absorption attenuation [53,54]. The fitting formula is shown below.

$$y = C_0 + C_1 e^{(-\frac{x}{\tau_1})} + C_2 e^{(-\frac{x}{\tau_2})} + C_3 e^{(-\frac{x}{\tau_3})}$$

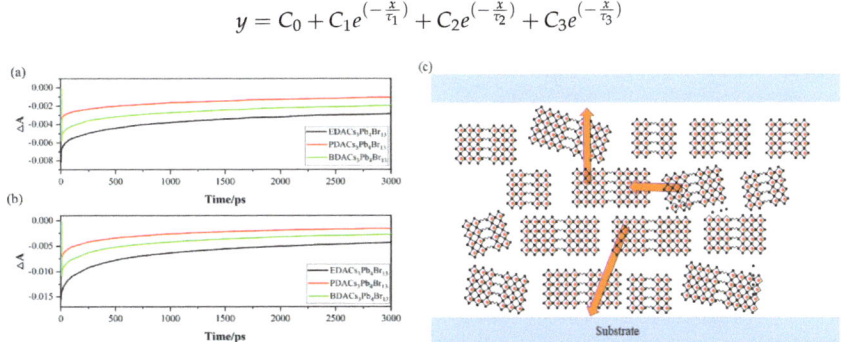

Figure 4. (**a**,**b**) Attenuation of transient absorption signals of EDA, PDA, and BDA excited from the top and bottom of the film. (**c**) Schematic diagram of energy transfer in perovskite thin films.

According to the time constants, the average carrier lifetime of different samples can be obtained. The average carrier lifetimes of EDA, PDA, and BDA are 2.11 ns, 2.37 ns, and 2.35 ns, respectively. The average carrier lifetime is calculated as follows:

$$\tau = \frac{C_1 \tau_1^2 + C_2 \tau_2^2 + C_3 \tau_3^2}{C_1 \tau_1 + C_2 \tau_2 + C_3 \tau_3}$$

The transient absorption signal intensity of the EDA samples is the highest of the three samples, and the second is the BDA samples. As shown in Figure 2a, the $n = 4$ exciton absorption peaks of the three samples are very close, which means that the $n = 4$ phase contents of the three perovskite films are close. As shown in Figure 4a, the enhancement of the transient absorption spectrum signal is accomplished in an extreme time. In such a short period of time, the trap has a limited effect on it. It can be inferred that the difference in the transient absorption spectrum signals of the three perovskite films mainly comes from ways other than the carrier generated by the $n = 4$ phase itself. We attribute the signal difference mainly to the energy transfer of perovskite from the small-n-value phase to the large-n-value phase. The transient absorption signal of three perovskite films was enhanced to the maximum within 1 ps, and the rising time of the signal was the same. It is obvious that the rising speed of the three samples is EDA > BDA > PDA. If the sample is excited from the top of the film, the rise time of the transient absorption spectrum signal is 0.5 ps longer than that from the bottom of the film. The EDA sample shows a faster and smoother energy-transfer process, which is mainly related to two factors. First, the EDA perovskite sample has a more favorable energy landscape for energy transfer, which can be seen from UV to Vis. The second is that EDA^{2+} has the shortest chain length among the three, which means that the distance between the inorganic layers of the low-dimensional phase in the EDA perovskite film sample is shorter than that between PDA and BDA. The shorter barrier reduces the quantum confinement effect and the dielectric confinement effect in the low-dimensional phase and weakens the effect of carriers being trapped in the inorganic layer. The PDA is supposed to have a stronger energy transfer than the BDA, but, in fact, it is the opposite. We think it is related to perovskite lattice distortion caused by PDA^{2+}. The lattice distortion in perovskite leads to the increase in the low-dimensional interphase barrier of perovskite. The increased barrier due to lattice distortion is even longer than the barrier when the organic cations are PDA^{2+} and BDA^{2+}. Therefore, the energy-transfer efficiency in PDA perovskite films is significantly reduced. Although EDA^{2+} and BDA^{2+}

organic spacer cations also cause perovskite lattice distortion, the degree of lattice distortion of perovskites is not as severe as that of PDA.

We can conclude a carrier transfer model of quasi-2D perovskite materials. After being excited by light, perovskite materials rapidly transition from the ground state to the excited state and produce a large number of hot carriers. Then, these hot carriers will move in the same n value phase and from small-n phases (wide band gap) to large-n phase (narrow band gap). The latter process is mainly related to two factors: the energy landscape of perovskite materials and the quantum confinement effect in perovskite materials. In several picoseconds after photoexcitation, the carriers also transfer rapidly to the perovskite phase, resulting in a population inversion and ASE signal in a very short time.

We have recorded the PL spectra of perovskite thin films at different locations of the pump light spot. It can be seen from Figure 5 that the PL peak value at the center of the pump light spot is generally stronger than that at the edge of the pump spot. This is related to the concentration of the carrier at the measurement spot. By comparing the PL peak at the edge of the area and the PL peak at the center of the area in different perovskite samples, we can obtain the result as shown in Figure 5. The attenuation of carrier concentration from the center of the spot to the edge of the EDA sample is the smallest, while that of PDA is the strongest. This means that the carrier transfer process in EDA samples is superior to that in BDA and PDA, which is in agreement with the conclusions obtained above.

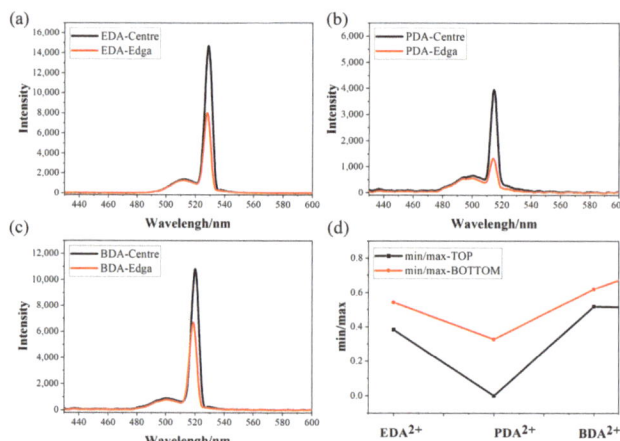

Figure 5. (**a**–**c**) The PL spectrums of EDA, PDA, and BDA are excited from the top and bottom of the films. (**d**) The ratio of the PL emission peak at the edge of the pump light spot to the central emission peak.

Although EDA perovskite has some advantages in fluorescence and energy transfer, it has a disadvantage that cannot be ignored compared to BDA and PDA. As seen in Figure 6, the transient absorption and fluorescence signals of EDA perovskite samples show a serious redshift (25.96 nm) after a while. This means that many of the original quasi-2D phases in EDA perovskite samples no longer have low-dimensional characteristics. The connection between the EDA^{2+} organic cation and the inorganic plate has been broken. Although the PDA sample and BDA sample also have a certain degree of redshift (PDA: 7.67 nm, BDA: 7.65 nm), compared with the EDA sample, the redshift of the PDA and BDA samples is not large. As shown in the picture, the degree of redshift of the PDA and BDA samples was the same. The result shows that the stability of the PDA and BDA samples is higher than that of the EDA sample.

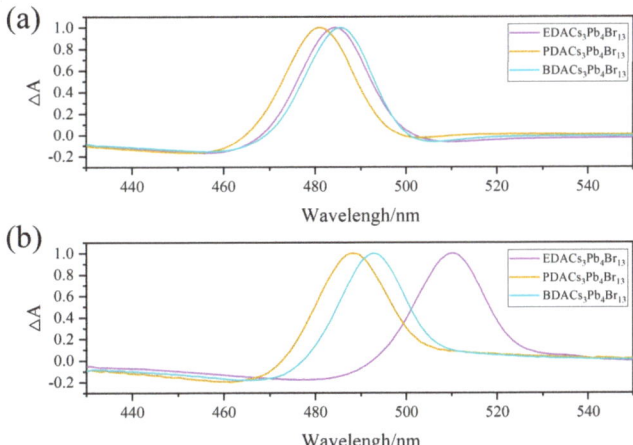

Figure 6. EDA, PDA, and BDA transient absorption signal after normalization: (**a**) initial stage of sample preparation, (**b**) 10 days after sample preparation.

3. Discussion

Based on the aforementioned experiments, the paper delves into the impact of organic spacer cations on perovskite. Notably, it was observed that shorter organic spacers result in reduced distances between inorganic layers. The organic cations follow the order of $EDA^{2+} < PDA^{2+} < BDA^{2+}$ in terms of length, with shorter lengths indicating smaller barriers and enhanced performance. However, barrier width is not the sole factor influencing perovskite performance. The two ends of the organic spacer extend into the connected inorganic layer, and the extent of this extension directly affects perovskite stability and lattice distortion.

As evidenced above, lattice distortion induced by PDA^{2+} is considerably greater than that caused by EDA^{2+} and BDA^{2+} [49]. Lattice distortion, in turn, alters the length of the Pb−Br bond, impacting the perovskite's band gap. Simultaneously, lattice distortion plays a role in modifying the barrier width to some degree. This phenomenon results in a larger band gap, diminished luminescence properties, and weaker energy transfer for PDA, as depicted in Figures 2 and 4.

The organic spacer EDA^{2+} has the shortest length, allowing it to penetrate the inorganic layer to a limited extent. Consequently, EDA exhibits a narrower barrier and superior performance. However, the shorter penetration distance compromises the stability of the connection between organic spacer cations and the inorganic layer in EDA, making it less stable compared to PDA and BDA, a trend consistent with Figure 6.

4. Materials and Methods

Perovskite thin film: The precursor solution was prepared according to the ratios of $CsBr:PbBr_2:EDABr_2 = 3:4:1$, $CsBr:PbBr_2:PDABr_2 = 3:4:1$, and $CsBr:PbBr_2:BDABr_2 = 3:4:1$. Dimethyl Sulfoxide (DMSO) was used as a solvent in the sample. Quartz flakes were cleaned with ethanol. The quartz sheet was then illuminated with ultraviolet light (UVO) for 15 min. The precursor solution was then spun onto the quartz sheet. Finally, the samples were annealed and encapsulated. The film was prepared using the spin-coating method and annealed at 70 °C.

Film characterization: The UV–Vis spectra of the films were obtained using the New Century T6 UV–Vis spectrophotometer. TRPL spectra were obtained using the Lifespec II instrument system. The transient absorption spectra of the thin films were measured by a self-built transient absorption spectroscopy system. The central wavelength of the laser was 800 nm, and the repetition frequency was 1 kHz. The pulse width of the laser used in the experiment was 40 fs. The system can achieve femtosecond time resolution.

5. Conclusions

In this study, quasi-2D perovskites featuring organic spacer cations of varying chain lengths were investigated. However, due to distortion effects, PDACs$_3$Pb$_4$Br$_{13}$ and BDACs$_3$Pb$_4$Br$_{13}$ samples were explicitly prepared for experiments. Notably, EDACs$_3$Pb$_4$Br$_{13}$, with the shortest organic isolation cation, exhibited weaker quantum-limiting effects and dielectric limiting compared to the other two samples. Remarkably, under identical excitation intensity, EDACs$_3$Pb$_4$Br$_{13}$ demonstrated a more vigorous luminous intensity. Additionally, in the short time following light excitation, EDACs$_3$Pb$_4$Br$_{13}$ exhibited faster carrier accumulation. If we consider the influence of organic spacer cation chain length on perovskite materials, the performance of the PDACs$_3$Pb$_4$Br$_{13}$ sample would be expected to surpass that of the BDACs$_3$Pb$_4$Br$_{13}$ sample. However, the distortion on the perovskite lattice caused by the PDA^{2+} organic spacer cation resulted in a decrease in the performance level of PDACs$_3$Pb$_4$Br$_{13}$ compared to its expected level. Furthermore, experimental results confirmed that the luminescence performance and energy transfer of PDACs$_3$Pb$_4$Br$_{13}$ were weaker than those of BDACs$_3$Pb$_4$Br$_{13}$. Despite the advantages exhibited by the EDACs$_3$Pb$_4$Br$_{13}$ perovskite sample over BDACs$_3$Pb$_4$Br$_{13}$, it was observed that the stability of the BDACs$_3$Pb$_4$Br$_{13}$ sample was significantly higher than that of the EDACs$_3$Pb$_4$Br$_{13}$. This suggests that, while EDACs$_3$Pb$_4$Br$_{13}$ may show favorable characteristics in certain aspects, the long-term stability of BDACs$_3$Pb$_4$Br$_{13}$ makes it a more promising candidate for practical applications.

Author Contributions: Conceptualization, Y.X. and L.S.; methodology, L.Z.; software, L.Z.; validation, L.Z. and Y.Z. (Yuan Zhang); resources, L.S. and X.G.; data curation, L.Z.; writing—original draft preparation, L.Z.; writing—review and editing, M.X. and Y.Z. (Yuan Zhang); visualization, M.X.; supervision, Y.Z. (Yong Zhang) and Y.W.; project administration, Y.X.; funding acquisition, Y.X. All authors have read and agreed to the published version of the manuscript.

Funding: This research was funded by The National Nature Science Foundation of China (62375074, 62075056) and The Nature Science Foundation of Hebei Province in China (F2021202055, F2022202035).

Data Availability Statement: The data presented in this study are available in this article.

Conflicts of Interest: Author Mingze Xia was employed by the company Cnooc Energy Technology and Services. The remaining authors declare that the research was conducted in the absence of any commercial or financial relationships that could be construed as potential conflicts of interest.

References

1. Stranks, S.D.; Snaith, H.J. Metal-halide perovskites for photovoltaic and light-emitting devices. *Nat. Nanotechnol.* **2015**, *10*, 391–402. [CrossRef] [PubMed]
2. Huang, P.; Kazim, S.; Wang, M.; Ahmad, S. Toward Phase Stability: Dion–Jacobson Layered Perovskite for Solar Cells. *ACS Energy Lett.* **2019**, *4*, 2960–2974. [CrossRef]
3. Zhao, W.; Dong, Q.; Zhang, J.; Wang, S.; Chen, M.; Zhao, C.; Hu, M.; Jin, S.; Padture, N.P.; Shi, Y. Asymmetric alkyl diamine based Dion–Jacobson low-dimensional perovskite solar cells with efficiency exceeding 15%. *J. Mater. Chem. A* **2020**, *8*, 9919–9926. [CrossRef]
4. He, T.; Li, S.; Jiang, Y.; Qin, C.; Cui, M.; Qiao, L.; Xu, H.; Yang, J.; Long, R.; Wang, H.; et al. Reduced-dimensional perovskite photovoltaics with homogeneous energy landscape. *Nature* **2020**, *11*, 1672. [CrossRef] [PubMed]
5. Fu, S.; Li, X.; Wan, L.; Zhang, W.; Song, W.; Fang, J. Effective Surface Treatment for High-Performance Inverted CsPbI2Br Perovskite Solar Cells with Efficiency of 15.92. *Nanomicro Lett.* **2020**, *12*, 170. [CrossRef] [PubMed]
6. Fu, P.; Liu, Y.; Yu, S.; Yin, H.; Yang, B.; Ahmad, S.; Guo, X.; Li, C. Dion-Jacobson and Ruddlesden-Popper double-phase 2D perovskites for solar cells. *Nano Energy* **2021**, *88*, 106249. [CrossRef]
7. Yang, J.; Yang, T.; Liu, D.; Zhang, Y.; Luo, T.; Lu, J.; Fang, J.; Wen, J.; Deng, Z.; Liu, S.; et al. Stable 2D Alternating Cation Perovskite Solar Cells with Power Conversion Efficiency >19% via Solvent Engineering. *Sol. RRL* **2021**, *5*, 2100286. [CrossRef]
8. Zhang, Y.; Wen, J.; Xu, Z.; Liu, D.; Yang, T.; Niu, T.; Luo, T.; Lu, J.; Fang, J.; Chang, X.; et al. Effective Phase-Alignment for 2D Halide Perovskites Incorporating Symmetric Diammonium Ion for Photovoltaics. *Adv. Sci.* **2021**, *8*, 2001433. [CrossRef]

9. Ren, G.; Yan, C.; Xiao, L.; Wu, X.; Peng, S.; Lin, W.; Tan, W.; Liu, Y.; Min, Y. Additive-Induced Film Morphology Evolution for Inverted Dion–Jacobson Quasi-Two-Dimensional Perovskite Solar Cells with Enhanced Performance. *ACS Appl. Energy Mater.* **2022**, *5*, 9837–9845. [CrossRef]
10. Yang, Z.; Liu, Z.; Ahmadi, V.; Chen, W.; Qi, Y. Recent Progress on Metal Halide Perovskite Solar Minimodules. *Sol. RRL* **2022**, *6*, 2100458. [CrossRef]
11. Li, M.; Gao, Q.; Liu, P.; Liao, Q.; Zhang, H.; Yao, J.; Hu, W.; Wu, Y.; Fu, H. Amplified Spontaneous Emission Based on 2D Ruddlesden-Popper Perovskites. *Adv. Funct. Mater.* **2018**, *28*, 1707006. [CrossRef]
12. Yu, M.; Yi, C.; Wang, N.; Zhang, L.; Zou, R.; Tong, Y.; Chen, H.; Cao, Y.; He, Y.; Wang, Y.; et al. Control of Barrier Width in Perovskite Multiple Quantum Wells for High Performance Green Light–Emitting Diodes. *Adv. Opt. Mater.* **2018**, *7*, 1801575. [CrossRef]
13. Chen, P.; Meng, Y.; Ahmadi, M.; Peng, Q.; Gao, C.; Xu, L.; Shao, M.; Xiong, Z.; Hu, B. Charge-transfer versus energy-transfer in quasi-2D perovskite light-emitting diodes. *Nano Energy* **2018**, *50*, 615–622. [CrossRef]
14. Das, S.; Gholipour, S.; Saliba, M. Perovskites for Laser and Detector Applications. *Energy Environ. Mater.* **2019**, *2*, 146–153. [CrossRef]
15. Li, Z.; Chen, Z.; Yang, Y.; Xue, Q.; Yip, H.L.; Cao, Y. Modulation of recombination zone position for quasi-two-dimensional blue perovskite light-emitting diodes with efficiency exceeding 5. *Nat. Commun.* **2019**, *10*, 1027. [CrossRef]
16. Zhu, L.; Liu, D.; Wang, J.; Wang, N. Large Organic Cations in Quasi-2D Perovskites for High-Performance Light-Emitting Diodes. *J. Phys. Chem. Lett.* **2020**, *11*, 8502–8510. [CrossRef]
17. De Giorgi, M.L.; Creti, A.; La-Placa, M.G.; Boix, P.P.; Bolink, H.J.; Lomascolo, M.; Anni, M. Amplified spontaneous emission in thin films of quasi-2D BA3MA3Pb5Br16 lead halide perovskites. *Nanoscale* **2021**, *13*, 8893–8900. [CrossRef]
18. Kar, S.; Jamaludin, N.F.; Yantara, N.; Mhaisalkar, S.G.; Leong, W.L. Recent advancements and perspectives on light management and high performance in perovskite light-emitting diodes. *Nanophotonics* **2021**, *10*, 2103–2143. [CrossRef]
19. Lian, Y.; Yang, Y.; He, L.; Yang, X.; Gao, J.; Qin, C.; Niu, L.; Yang, X. Enhancing the Luminance Efficiency of Formamidinium-Based Dion-Jacobson Perovskite Light-Emitting Diodes via Compositional Engineering. *ACS Appl. Mater. Interfaces* **2021**, *14*, 1659–1669. [CrossRef]
20. Qin, X.; Liu, F.; Leung, T.L.; Sun, W.; Chan, C.C.S.; Wong, K.S.; Kanižaj, L.; Popović, J.; Djurišić, A.B. Compositional optimization of mixed cation Dion–Jacobson perovskites for efficient green light emission. *J. Mater. Chem. C* **2022**, *10*, 108–114. [CrossRef]
21. Lian, Y.; Yang, Y.; Gao, J.; Qin, C. Efficient Dion-Jacobson perovskite light-emitting diodes via mixed cation engineering. *Opt. Lett.* **2022**, *47*, 657–660. [CrossRef] [PubMed]
22. Zhong, Y.; Liu, G.; Su, Y.; Sheng, W.; Gong, L.; Zhang, J.; Tan, L.; Chen, Y. Diammonium molecular configuration-induced regulation of crystal orientation and carrier dynamics for highly efficient and stable 2D/3D perovskite solar cells. *Angew. Chem. Int. Ed.* **2022**, *61*, e202114588. [CrossRef] [PubMed]
23. Yu, H.; Xie, Y.; Zhang, J.; Duan, J.; Chen, X.; Liang, Y.; Wang, K.; Xu, L. Thermal and humidity stability of mixed spacer cations 2D perovskite solar cells. *Adv. Sci.* **2021**, *8*, 2004510. [CrossRef] [PubMed]
24. Xu, J.; Chen, J.; Chen, S.; Gao, H.; Li, Y.; Jiang, Z.; Zhang, Y.; Wang, X.; Zhu, X.; Xu, B. Organic spacer engineering of ruddlesden-popper perovskite materials toward efficient and stable solar cells. *Chem. Eng. J.* **2023**, *453*, 139790. [CrossRef]
25. Yang, T.; Ma, C.; Cai, W.; Wang, S.; Wu, Y.; Feng, J.; Wu, N.; Li, H.; Huang, W.; Ding, Z. Amidino-based Dion-Jacobson 2D perovskite for efficient and stable 2D/3D heterostructure perovskite solar cells. *Joule* **2023**, *7*, 574–586. [CrossRef]
26. Ngai, K.H.; Wei, Q.; Chen, Z.; Guo, X.; Qin, M.; Xie, F.; Chan, C.C.S.; Xing, G.; Lu, X.; Chen, J. Enhanced electrochemical stability by alkyldiammonium in Dion–Jacobson perovskite toward ultrastable light-emitting diodes. *Adv. Opt. Mater.* **2021**, *9*, 2100243. [CrossRef]
27. Lei, Y.; Li, Z.; Wang, H.; Wang, Q.; Peng, G.; Xu, Y.; Zhang, H.; Wang, G.; Ding, L.; Jin, Z. Manipulate energy transport via fluorinated spacers towards record-efficiency 2D Dion-Jacobson CsPbI3 solar cells. *Sci. Bull.* **2022**, *67*, 1352–1361. [CrossRef] [PubMed]
28. Ke, W.; Mao, L.; Stoumpos, C.C.; Hoffman, J.; Spanopoulos, I.; Mohite, A.D.; Kanatzidis, M.G. Compositional and solvent engineering in Dion–Jacobson 2D perovskites boosts solar cell efficiency and stability. *Adv. Energy Mater.* **2019**, *9*, 1803384. [CrossRef]
29. Wu, H.; Lian, X.; Tian, S.; Zhang, Y.; Qin, M.; Zhang, Y.; Wang, F.; Lu, X.; Wu, G.; Chen, H. Additive-Assisted Hot-Casting Free Fabrication of Dion–Jacobson 2D Perovskite Solar Cell with Efficiency Beyond 16%. *Sol. RRL* **2020**, *4*, 2000087. [CrossRef]
30. Mitzi, D.B. Templating and structural engineering in organic–inorganic perovskites. *J. Chem. Soc. Dalton Trans.* **2001**, *1*, 1–12. [CrossRef]
31. Hu, T.; Smith, M.D.; Dohner, E.R.; Sher, M.-J.; Wu, X.; Trinh, M.T.; Fisher, A.; Corbett, J.; Zhu, X.Y.; Karunadasa, H.I. Mechanism for broadband white-light emission from two-dimensional (110) hybrid perovskites. *J. Phys. Chem. Lett.* **2016**, *7*, 2258–2263. [CrossRef] [PubMed]
32. Tan, M.; Wang, S.; Rao, F.; Yang, S.; Wang, F. Pressures tuning the band gap of organic–inorganic trihalide perovskites (MAPbBr 3): A first-principles study. *J. Electron. Mater.* **2018**, *47*, 7204–7211. [CrossRef]
33. Mao, L.; Guo, P.; Kepenekian, M.; Hadar, I.; Katan, C.; Even, J.; Schaller, R.D.; Stoumpos, C.C.; Kanatzidis, M.G. Structural diversity in white-light-emitting hybrid lead bromide perovskites. *J. Am. Chem. Soc.* **2018**, *140*, 13078–13088. [CrossRef] [PubMed]

34. Saidaminov, M.I.; Mohammed, O.F.; Bakr, O.M. Low-dimensional-networked metal halide perovskites: The next big thing. *ACS Energy Lett.* **2017**, *2*, 889–896. [CrossRef]
35. Ahmad, S.; Hanmandlu, C.; Kanaujia, P.K.; Prakash, G.V. Direct deposition strategy for highly ordered inorganic organic perovskite thin films and their optoelectronic applications. *Opt. Mater. Express* **2014**, *4*, 1313–1323. [CrossRef]
36. Hong, X.; Ishihara, T.; Nurmikko, A.V. Dielectric confinement effect on excitons in PbI 4-based layered semiconductors. *Phys. Rev. B* **1992**, *45*, 6961. [CrossRef] [PubMed]
37. Muljarov, E.A.; Tikhodeev, S.G.; Gippius, N.A.; Ishihara, T. Excitons in self-organized semiconductor/insulator superlattices: PbI-based perovskite compounds. *Phys. Rev. B* **1995**, *51*, 14370. [CrossRef] [PubMed]
38. Chen, Y.; Sun, Y.; Peng, J.; Zhang, W.; Su, X.; Zheng, K.; Pullerits, T.; Liang, Z. Tailoring organic cation of 2D air-stable organometal halide perovskites for highly efficient planar solar cells. *Adv. Energy Mater.* **2017**, *7*, 1700162. [CrossRef]
39. Guo, X.; Gao, Y.; Wei, Q.; Ho Ngai, K.; Qin, M.; Lu, X.; Xing, G.; Shi, T.; Xie, W.; Xu, J. Suppressed Phase Segregation in High-Humidity-Processed Dion–Jacobson Perovskite Solar Cells Toward High Efficiency and Stability. *Sol. RRL* **2021**, *5*, 2100555. [CrossRef]
40. Proppe, A.H.; Quintero-Bermudez, R.; Tan, H.; Voznyy, O.; Kelley, S.O.; Sargent, E.H. Synthetic control over quantum well width distribution and carrier migration in low-dimensional perovskite photovoltaics. *J. Am. Chem. Soc.* **2018**, *140*, 2890–2896. [CrossRef]
41. Li, X.; Hoffman, J.; Ke, W.; Chen, M.; Tsai, H.; Nie, W.; Mohite, A.D.; Kepenekian, M.; Katan, C.; Even, J. Two-dimensional halide perovskites incorporating straight chain symmetric diammonium ions, $(NH_3C_mH_{2m}NH_3)(CH_3NH_3)_{n-1}Pb_nI_{3n+1}$ (m = 4–9; n = 1–4). *J. Am. Chem. Soc.* **2018**, *140*, 12226–12238. [CrossRef] [PubMed]
42. Ma, C.; Shen, D.; Ng, T.W.; Lo, M.F.; Lee, C.S. 2D perovskites with short interlayer distance for high-performance solar cell application. *Adv. Mater.* **2018**, *30*, 1800710. [CrossRef] [PubMed]
43. Ahmad, S.; Fu, P.; Yu, S.; Yang, Q.; Liu, X.; Wang, X.; Wang, X.; Guo, X.; Li, C. Dion-Jacobson phase 2D layered perovskites for solar cells with ultrahigh stability. *Joule* **2019**, *3*, 794–806. [CrossRef]
44. Zheng, Y.; Niu, T.; Qiu, J.; Chao, L.; Li, B.; Yang, Y.; Li, Q.; Lin, C.; Gao, X.; Zhang, C. Oriented and uniform distribution of Dion–Jacobson phase perovskites controlled by quantum well barrier thickness. *Sol. RRL* **2019**, *3*, 1900090. [CrossRef]
45. Sourisseau, S.; Louvain, N.; Bi, W.; Mercier, N.; Rondeau, D.; Boucher, F.; Buzaré, J.-Y.; Legein, C. Reduced band gap hybrid perovskites resulting from combined hydrogen and halogen bonding at the organic− inorganic interface. *Chem. Mater.* **2007**, *19*, 600–607. [CrossRef]
46. Gan, L.; Li, J.; Fang, Z.; He, H.; Ye, Z. Effects of organic cation length on exciton recombination in two-dimensional layered lead iodide hybrid perovskite crystals. *J. Phys. Chem. Lett.* **2017**, *8*, 5177–5183. [CrossRef]
47. Blancon, J.C.; Tsai, H.; Nie, W.; Stoumpos, C.C.; Pedesseau, L.; Katan, C.; Kepenekian, M.; Soe, C.M.M.; Appavoo, K.; Sfeir, M.Y. Extremely efficient internal exciton dissociation through edge states in layered 2D perovskites. *Science* **2017**, *355*, 1288–1292. [CrossRef]
48. Han, Y.; Li, Y.; Wang, Y.; Cao, G.; Yue, S.; Zhang, L.; Cui, B.B.; Chen, Q. From distortion to disconnection: Linear alkyl diammonium cations tune structure and photoluminescence of lead bromide perovskites. *Adv. Opt. Mater.* **2020**, *8*, 1902051. [CrossRef]
49. Ajayakumar, A.; Muthu, C.; Dev, A.V.; Pious, J.K.; Vijayakumar, C. Two-Dimensional Halide Perovskites: Approaches to Improve Optoelectronic Properties. *Chem.—Asian J.* **2022**, *17*, e202101075. [CrossRef]
50. Queloz, V.I.E.; Bouduban, M.E.F.; García-Benito, I.; Fedorovskiy, A.; Orlandi, S.; Cavazzini, M.; Pozzi, G.; Trivedi, H.; Lupascu, D.C.; Beljonne, D. Spatial charge separation as the origin of anomalous stark effect in fluorous 2d hybrid perovskites. *Adv. Funct. Mater.* **2020**, *30*, 2000228. [CrossRef]
51. Walters, G.; Wei, M.; Voznyy, O.; Quintero-Bermudez, R.; Kiani, A.; Smilgies, D.M.; Munir, R.; Amassian, A.; Hoogland, S.; Sargent, E. The quantum-confined Stark effect in layered hybrid perovskites mediated by orientational polarizability of confined dipoles. *Nat. Commun.* **2018**, *9*, 4214. [CrossRef] [PubMed]
52. Sharma, D.K.; Hirata, S.; Biju, V.; Vacha, M. Stark effect and environment-induced modulation of emission in single halide perovskite nanocrystals. *ACS Nano* **2019**, *13*, 624–632. [CrossRef] [PubMed]
53. Manser, J.S.; Kamat, P.V. Band filling with free charge carriers in organometal halide perovskites. *Nat. Photonics* **2014**, *8*, 737–743. [CrossRef]
54. Konidakis, I.; Maksudov, T.; Serpetzoglou, E.; Kakavelakis, G.; Kymakis, E.; Stratakis, E. Improved Charge Carrier Dynamics of $CH_3NH_3PbI_3$ Perovskite Films Synthesized by Means of Laser-Assisted Crystallization. *ACS Appl. Energy Mater.* **2018**, *1*, 5101–5111. [CrossRef]

Disclaimer/Publisher's Note: The statements, opinions and data contained in all publications are solely those of the individual author(s) and contributor(s) and not of MDPI and/or the editor(s). MDPI and/or the editor(s) disclaim responsibility for any injury to people or property resulting from any ideas, methods, instructions or products referred to in the content.

Article

Growth, Spectroscopic Characterization and Continuous-Wave Laser Operation of Er,Yb:GdMgB$_5$O$_{10}$ Crystal

Konstantin N. Gorbachenya [1], Elena A. Volkova [2,*], Victor V. Maltsev [2,*], Victor E. Kisel [1], Diana D. Mitina [2], Elizaveta V. Koporulina [2], Nikolai N. Kuzmin [3], Ekaterina I. Marchenko [2] and Vladimir L. Kosorukov [2]

1. Center for Optical Materials and Technologies, Belarusian National Technical University, Nezavisimosti Ave., 65, 220013 Minsk, Belarus; gorby@bntu.by (K.N.G.)
2. Department of Crystallography and Crystal Chemistry, Faculty of Geology, Moscow State University, 119234 Moscow, Russia; e_koporulina@mail.ru (E.V.K.)
3. Institute of Spectroscopy, Russian Academy of Sciences, 108840 Moscow, Russia
* Correspondence: volkova@geol.msu.ru (E.A.V.); maltsev@geol.msu.ru (V.V.M.)

Abstract: A transparent Er^{3+},Yb^{3+}:GdMgB$_5$O$_{10}$ single crystal with dimensions up to 24 × 15 × 12 mm was grown successfully by the high-temperature solution growth on dipped seeds technique from K$_2$Mo$_3$O$_{10}$-based solvent. The grown crystal was characterized using PXRD, DSC and ATR techniques. Differential scanning calorimetry measurements and SEM analysis of the heat-treated solids revealed Er,Yb:GdMgB$_5$O$_{10}$ to be an incongruent melting compound with an onset point of 1087 °C. The absorption edge of the Er,Yb:GMBO sample is located in the region of 245 nm, which approximates a value of 4.8 eV. Absorption and emission spectra, and luminescence kinetics, were studied. The energy transfer efficiency from ytterbium to erbium ions was determined. The laser operation in continuous-wave mode was realized and output characteristics were measured. The maximal output power of 0.15 W with a slope efficiency of 11% was obtained at 1568 nm.

Keywords: erbium; ytterbium; pentaborate crystal; growth; electronic band; differential scanning calorimetry; spectroscopy; laser operation

Citation: Gorbachenya, K.N.; Volkova, E.A.; Maltsev, V.V.; Kisel, V.E.; Mitina, D.D.; Koporulina, E.V.; Kuzmin, N.N.; Marchenko, E.I.; Kosorukov, V.L. Growth, Spectroscopic Characterization and Continuous-Wave Laser Operation of Er,Yb:GdMgB$_5$O$_{10}$Crystal. *Inorganics* **2024**, *12*, 240. https://doi.org/10.3390/inorganics12090240

Academic Editor: Binbin Chen

Received: 23 July 2024
Revised: 26 August 2024
Accepted: 28 August 2024
Published: 31 August 2024

Copyright: © 2024 by the authors. Licensee MDPI, Basel, Switzerland. This article is an open access article distributed under the terms and conditions of the Creative Commons Attribution (CC BY) license (https://creativecommons.org/licenses/by/4.0/).

1. Introduction

Borate compounds have been extensively studied over the past few decades due to their remarkable structural flexibility and potential applications as laser, nonlinear, scintillation, magnetic and phosphor materials, etc., [1]. The variety of functional properties of borates is based on the diversity of their structural types: boron atoms can be bonded with three or four oxygen atoms, forming planar/non-planar triangular BO$_3$ or tetrahedral BO$_4$ fundamental structural units, respectively. These B-O building groups can be connected to each other by common corners or edges, forming different B-O clusters [2]. Nowadays, rare-earth metal borates are very attractive objects for the research community because of their nonlinear optical and laser applications.

Laser radiation in the 1.5–1.6 μm spectral range is extensively used in range finding, optical locating and telecommunications applications, mainly because of its eye safety, weak absorption in the atmosphere and low dispersion and absorption of quartz fibers. This radiation can be obtained using solid-state lasers based on gain media doped with trivalent erbium ions (transition $^4I_{13/2} \rightarrow ^4I_{15/2}$). However, the main disadvantage of the Er^{3+} ions is low absorption in the spectral range of InGaAs laser diode emissions (near 1 μm), which limits pumping efficiency. Ytterbium ions are a good sensitizer due to the broad absorption band near 1 μm and the large overlap between Yb^{3+} emissions and Er^{3+} absorption, which allows for resonant energy transfer from ytterbium to erbium ions [3].

Recently, the spectroscopic and laser properties of different gain media have been investigated. Among them, phosphate glasses co-doped with Er,Yb ions are the most widely used, since Er,Yb-phosphate glasses are characterized by spectroscopic properties suitable for efficient laser operation (energy transfer from Yb^{3+} to Er^{3+} ions of 90% with a long lifetime of the erbium upper laser level $^4I_{13/2}$ of 7–8 ms and a short lifetime of the $^4I_{11/2}$ energy level of 2–3 µs) [4]. However, phosphate glass exhibits poor thermo-mechanical properties (a thermal conductivity of $0.85\ W \times m^{-1} \times K^{-1}$) [5], which limits the average output power of Er,Yb:glass lasers in continuous-wave and Q-switched regimes of operation due to the thermal effects.

Crystalline laser hosts are characterized by significantly higher thermal conductivity values than glasses [6,7]. Currently, many crystalline hosts have been investigated for Er,Yb lasers—aluminates, silicates, vanadates, and tungstates [8,9]—but spectroscopic properties of these crystals do not fully meet the requirements for achieving an efficient laser operation. Nowadays, oxoborate crystals are the most important Er,Yb-codoped crystalline laser materials because they possess not only high thermal conductivity but also the necessary spectroscopic properties mentioned above [10,11]. The most efficient laser operation in continuous-wave mode has been demonstrated for huntite-type Er,Yb:$RAl_3(BO_3)_4$ (R = Y, Gd, Lu) [12–14] and pentaborate Er,Yb:$RMgB_5O_{10}$ [15–18] (R = Y, Gd, La) crystals. The pentaborate crystals demonstrate relatively high thermal conductivity and good spectroscopic properties [16]. The highest output power of 0.61 W and slope efficiency of 23% was realized for Er,Yb:$LaMgB_5O_{10}$ [18]. We have recently exhibited a continuous-wave Er,Yb:$YMgB_5O_{10}$ laser with an output power of 0.2 W at 1570 nm [15]. $GdMgB_5O_{10}$ (GMBO) crystal also proved to be a good choice for both Er,Yb-codoping and single Yb-doping. The maximal output power of 0.22 W with a slope efficiency of 14% at 1569 nm was demonstrated for Er,Yb:$GdMgB_5O_{10}$ [19], while up to 5.35 W and 56% of maximal output power and slope efficiency, respectively, were demonstrated at 1060.8 nm for Yb:$GdMgB_5O_{10}$ [20].

In the present work, a comprehensive characterization including growth technique, thermal stability, IR spectroscopy, bandgap transmission spectra, and spectroscopic and laser properties of as-grown Er,Yb:$GdMgB_5O_{10}$ single crystals was carried out.

2. Results and Discussion

2.1. Growth Technique and Structure Characterization

A transparent macrodefect-free Er,Yb-codoped $GdMgB_5O_{10}$ single crystal with a size of 24 × 15 × 12 mm was grown by the high-temperature solution growth on dipped seeds (HT-SGDS) technique (Figure 1). The saturation temperature value obtained was about ~950 °C.

Figure 1. As-grown Er,Yb:$GdMgB_5O_{10}$ single crystal.

The Le Bail fitting was performed to confirm the structural similarity of Er,Yb:GMBO to rare-earth magnesium pentaborates. To perform the Le Bail method, the crystal symmetry and lattice parameters for $GdMgB_5O_{10}$ (database code: ICSD 157426) were used as input data. Powder X-ray diffraction (PXRD) refinements showed good agreement with the theoretical profile. Nevertheless, pattern matching revealed an impurity phase, SiO_2, which appeared after grinding in the agate mortar (Figure 2).

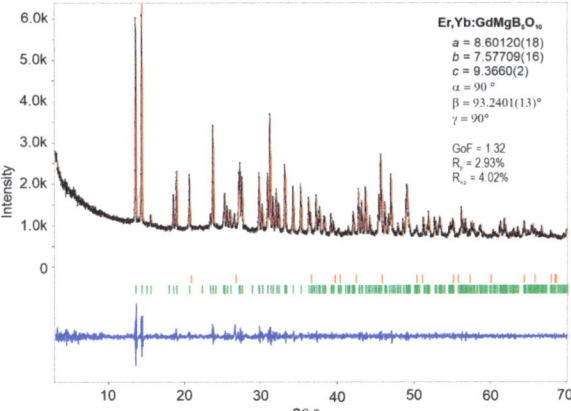

Figure 2. Diffraction profiles showing observed (black crosses) and calculated (red continuous line) profiles, and the difference curve (blue continuous line) between observed and calculated spectra. Green and orange vertical markers represent Bragg reflections corresponding to the main phase of $GdMgB_5O_{10}$ (ICSD 157426) and the impurity phase of SiO_2 (COD 1526860), respectively.

The thermal behavior of the Er,Yb:GMBO compound was determined by differential scanning calorimetry (DSC). A fragment of calorimetric data in the temperature range 600–1200 °C is shown in Figure 3. The heating curve is characterized by a sharp endothermic peak with an onset temperature of ~1087 °C. However, no exothermic peak is observed on the cooling curve, indicating the incongruent melting of the compound under investigation.

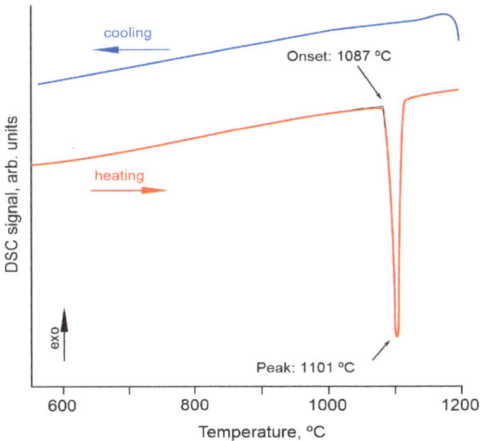

Figure 3. DSC curves of Er,Yb:$GdMgB_5O_{10}$ crystal in the temperature range of 600–1200 °C.

Similar thermal behavior has been previously described for $TmMgB_5O_{10}$ and $YMgB_5O_{10}$, which undergo incongruent melting to form RBO_3 and $Mg_2B_2O_{10}$ phases [21]. Spontaneous Er,Yb:GMBO crystals from the crystallized (after cooling) molten bath were annealed in an alundum crucible at 1050 °C and characterized using the SEM/EDXS technique. The residue in the crucible after heat treatment was a white, translucent, dense mass containing mostly partially melted pentaborate crystals (Figure 4). SEM/EDXS data revealed the existence of two phases other than Er,Yb:GMBO.

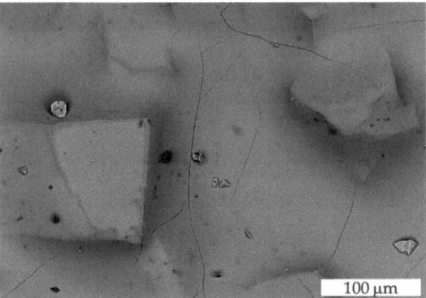

Figure 4. SEM image of the Er,Yb:GMBO spontaneous crystals heat-treated at 1050 °C.

The distribution patterns of the elements in the X-ray characteristic radiation show that the bright phase formed is enriched in the rare-earth component and the darker one is enriched in magnesium (Figure 5). It is impossible to determine the distribution of boron in the newly formed phases from the SEM data, but it can be assumed that the first phase is represented by rare-earth orthoborate RBO_3 (due to the high stability of this phase in a wide temperature range), and the second by the magnesium phase. Considering previous studies on the thermal properties of Tm and YMg pentaborates, the thermally induced process can be described by the following reaction:

$$2RMgB_5O_{10} \rightarrow 2RBO_3 + Mg_2B_2O_5 + 3B_2O_3, \text{ where } R = \text{Er, Yb, and Gd.}$$

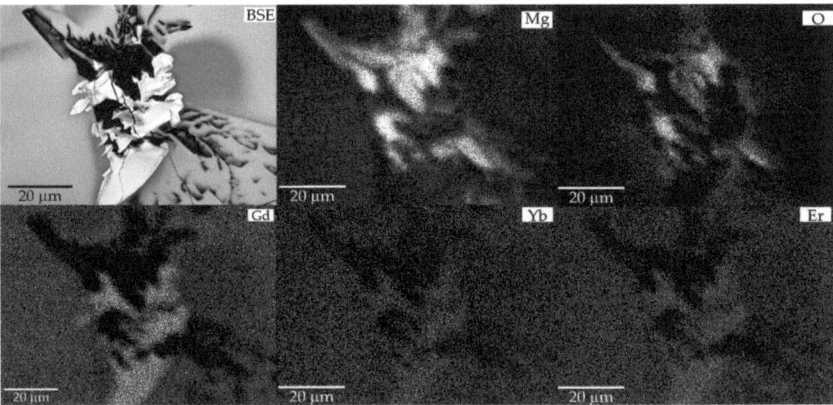

Figure 5. SEM/EDXS patterns of the decomposition products after annealing Er,Yb:GMBO spontaneous crystals.

Figure 6 shows the ATR spectrum of Er,Yb:GdMgB$_5$O$_{10}$. Factor-group analysis for $RMgB_5O_{10}$ compounds was performed in $RMgB_5O_{10}$ compounds [21]. According to it, the GdMgB$_5$O$_{10}$ compound exhibits 201 optically active modes, of which 99 are IR active. The ATR spectrum contains 40 modes. The smaller number of observed modes is explained by the fact that they cannot be resolved due to the large number of lattice vibrations with close frequencies, and that below 250 cm^{-1} there is strong absorption, associated with intense modes [21].

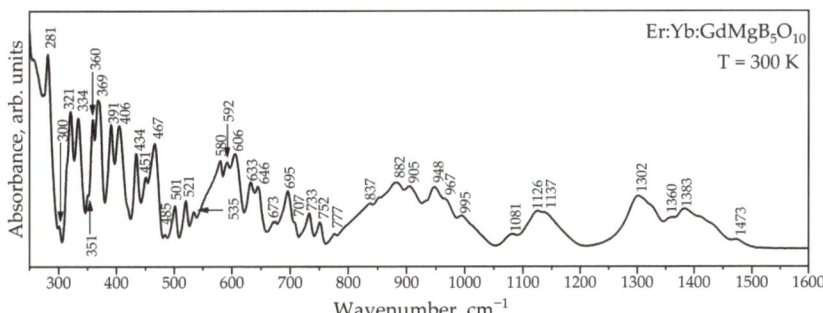

Figure 6. ATR spectrum of Er:Yb:GdMgB$_5$O$_{10}$.

Figure 7a demonstrates the transmission spectrum of Er,Yb:GdMgB$_5$O$_{10}$ crystal in the range of 220–300 nm. The absorption edge of the Er,Yb:GMBO sample is located in the region of 245 nm, which is about 30 nm larger than for YMgB$_5$O$_{10}$ [22]. The optical band gap (E_g) obtained from transmission spectra approximates a value of 4.8 eV (Figure 7b) according to methodology from [23]. The formulas for indirect gap semiconductors were used to calculate the band gap.

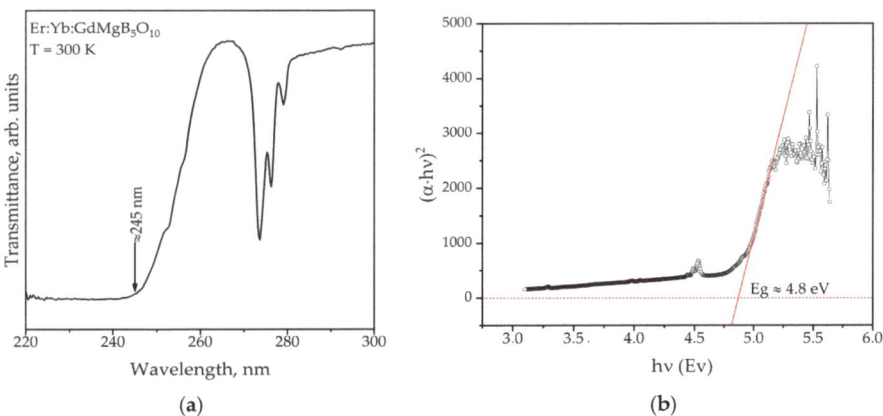

Figure 7. Transmission spectrum (**a**) and related Tauc plot (**b**) of the Er:Yb:GdMgB$_5$O$_{10}$ crystal.

2.2. Spectroscopy

The room-temperature polarized absorption cross-section spectra of the Er,Yb:GMBO crystal in the 850–1100 nm spectral range are shown in Figure 8. The peak absorption cross-sections around 976 nm (the wavelength is close to the emission wavelengths of commercial InGaAs laser diodes) corresponding to the $^2F_{7/2} \rightarrow {}^2F_{5/2}$ transitions of Yb^{3+} ions are about 1.5×10^{-20} cm^2 for $E//N_g$ polarization. Therefore, the polarization of the pump beam that matches the N_g axis of the crystal is preferred. The FWHM near 976 nm is ~2 nm, which imposes additional requirements for the wavelength stabilization of laser diodes.

Figure 9 shows the polarized absorption cross-section of the Er,Yb:GMBO crystal at room temperature near 1.5 µm. Structured spectra with several narrow bands related to the $^4I_{15/2} \rightarrow {}^4I_{13/2}$ transition of Er^{3+} ions are observed in the 1400–1650 nm spectral range. Multiple peaks in the absorption cross-section spectra correspond to transitions between the stark sublevels of the lower and upper multiplets. One can note the similarity with the data presented for Er,Yb:YMBO in [15]. The maximum absorption cross-section does not exceed 1.1×10^{-20} cm^2 at the wavelength of 1530 nm.

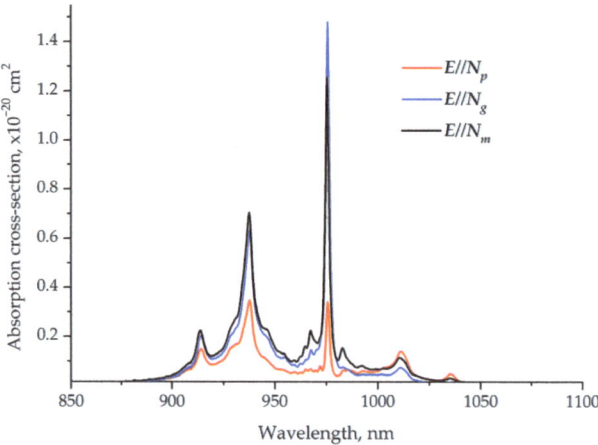

Figure 8. Polarized room-temperature absorption cross-section spectra of Er,Yb:GMBO crystal near 1 μm.

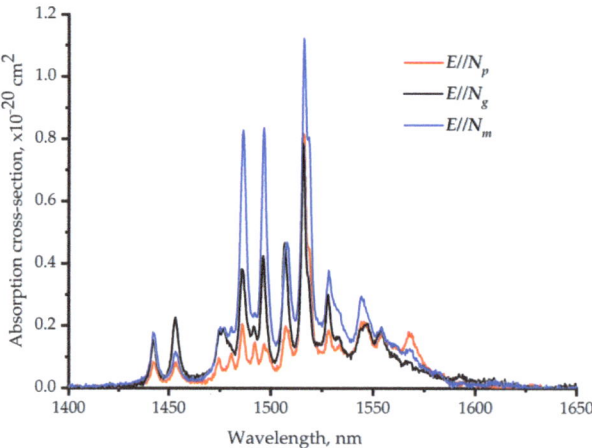

Figure 9. Polarized room-temperature absorption cross-section spectra of the Er,Yb:GMBO crystal near 1.5 μm.

The decay curve of the 1.5 μm emission is shown in Figure 10. It was approximated by a single exponential function, and the $^4I_{13/2}$ energy level of Er^{3+} was found to be 430 ± 20 μs. We can note that the influence of reabsorption is not significant for the $^4I_{13/2}$ erbium energy level measurements. The emission lifetime was calculated using the Judd–Ofelt method and amounted to 6.30 ms (the corresponding data are given in [16]). Thus, the quantum yield of luminescence does not exceed 7%, which is comparable to values demonstrated previously for other oxoborate crystals with high phonon energy [13].

To calculate energy transfer efficiency from ytterbium to erbium according to Equation (1), the lifetime of the $^2F_{5/2}$ energy level of Yb^{3+} in Yb single-doped crystal and the Er,Yb-codoped one was measured. The dependence of the measured $^2F_{5/2}$-energy-level lifetime of Yb^{3+} on ytterbium content using Yb:GMBO powder in glycerin is presented in Figure 11.

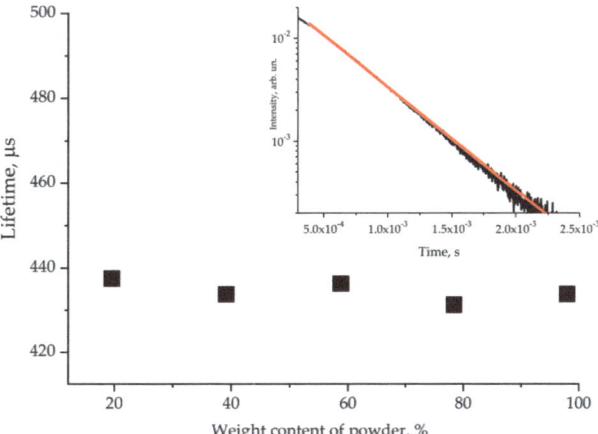

Figure 10. The lifetime of the erbium $^4I_{13/2}$ energy level measured using Er,Yb:GMBO crystalline powder in glycerin. The inset shows luminescence decay kinetics near 1.5 μm.

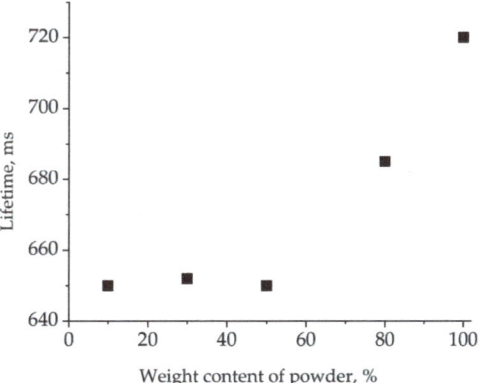

Figure 11. Lifetime of the $^2F_{5/2}$ Yb^{3+} energy level measured using Yb:GMBO crystalline powder in glycerin.

The measured lifetime decreased with decreasing powder content in the suspension. Starting from a certain powder content, the lifetime did not change despite further dilution, suggesting a negligible influence of reabsorption. The lifetime of the Yb^{3+} $^2F_{5/2}$ energy level in the Yb(1 at.%):GMBO crystal was 650 ± 30 μs. The measurement of the Yb^{3+}$^2F_{5/2}$-energy-level lifetime of Er,Yb:GMBO was similar to above and the $^2F_{5/2}$-energy-level lifetime of Er,Yb:GMBO crystal was determined to be 60 ± 5 μs. Thus, the calculated energy transfer efficiency in the Er,Yb:GMBO crystal doped with 2 at.% Er^{3+} and 11 at.% Yb^{3+} was close to 90%. These results show that the almost all of the absorbed energy may be efficiently transferred from the $^2F_{5/2}$ energy level of the Yb^{3+} ion to the $^4I_{11/2}$ energy level of the Er^{3+} ion by a non-radiative resonant energy transfer process. It is also worth noting that the energy transfer efficiency in Er,Yb:GMBO crystals is similar to that of Er,Yb:RAl$_3$(BO$_3$)$_4$ and Er,Yb:YMgB$_5$O$_{10}$ [15].

The stimulated emission cross-section spectra of Er,Yb:GMBO crystal in the 1400–1650 nm range are presented in Figure 12. The emission cross-section is 0.7×10^{-20} cm^2 at a laser wavelength of 1568 nm for the polarization of $E//N_p$. The inset shows the gain cross-section spectra of the Er,Yb:GMBO crystal for different inversion parameters β ranging from 0.2 to 0.8 in the 1480–1600 nm spectral range for $E//N_p$ polarization.

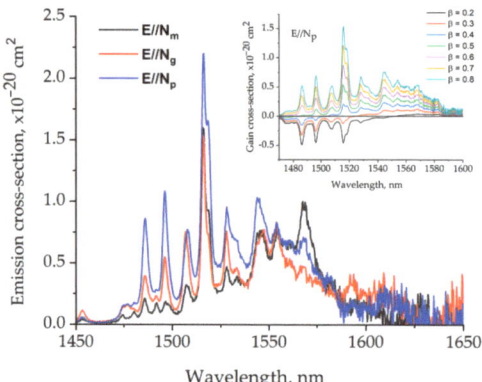

Figure 12. The stimulated emission cross-section spectra of the Er,Yb:GMBO crystal. The inset shows the gain cross-section spectra of the Er,Yb:GMBO crystal.

2.3. Laser Operation

The continuous-wave Er,Yb:GMBO laser operation was realized for output couplers (OCs) with transmissions of 1%, 2% and 5%. The maximal output power and slope efficiency was obtained for OCs with a transmission of 2% at the laser wavelength. The input–output characteristics of a continuous-wave Er,Yb:GMBO laser with a 2% OC is presented in Figure 13. The maximum output power of 0.15 W at 1568 nm with a slope efficiency of 11% was achieved at an absorbed pump power of 3.3 W. Degradation of the output characteristics was observed with increasing incident power, which may be due to the influence of thermal lensing. The laser threshold was approximately 1.8 W of absorbed pump power. The laser radiation was linearly polarized ($E//N_p$). The spatial profile of the output beam was TEM$_{00}$ mode with $M^2 < 1.2$ during all laser operations (inset in Figure 13).

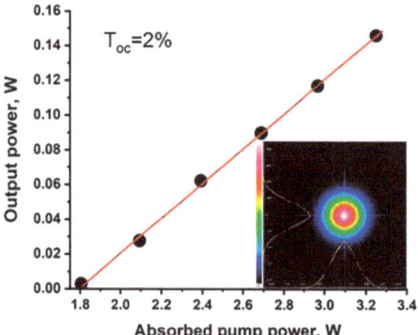

Figure 13. The input–output characteristics of Er,Yb:GMBO laser; the inset shows the spatial profile of the output beam.

3. Materials and Methods

(Er,Yb):GdMgB$_5$O$_{10}$ bulk crystal was grown by the HT-SGDS technique from K$_2$Mo$_3$O$_{10}$ solvent in the temperature range 870–830 °C. Previously, experiments on GMBO spontaneous crystallization were carried out to estimate acceptable fluxed melt composition, taking into account the peculiarities of phase formation in other RMBO-K$_2$Mo$_3$O$_{10}$ systems [24]. Based on the results obtained, a complex system of the composition 20 wt.% Er,Yb:GMBO–80 wt.% K$_2$Mo$_3$O$_{10}$ was applied in growth experiments. The R_2O_3/MgO/B$_2$O$_3$ (where R = Yb, Er, Gd) ratio was set according to the pentaborate formula (1:2:5 in molar fractions). Er$_2$O$_3$ (99.96%), Yb$_2$O$_3$ (99.96%), Gd$_2$O$_3$ (99.96%), MgO (A.C.S. grade, by Aldrich), and B$_2$O$_3$

(A.C.S. grade, by Alfa Aesar) were used as crystal-forming agents, which were weighed according to the composition of $Er_{0.02}Yb_{0.11}Gd_{0.87}MgB_5O_{10}$. The solvent, $K_2Mo_3O_{10}$, was prepared from a mixture of K_2MoO_4 (A.C.S. grade, Chimkraft, Kaliningrad, Russia) and MoO_3 (A.C.S. grade, Aldrich), compounded in a molar ratio of 1:2.

The starting chemicals were carefully ground, mixed, and in the 250 mL platinum crucible placed into the furnace at a position that could be adjusted for an optimum temperature gradient. Then, the initial load was heated to a maximum temperature of 100–150 °C above the expected saturation temperature (T_{sat}). After solution homogenization for 24 h, the system was cooled to 5–10 °C above the expected T_{sat}. The saturation point was then precisely confirmed by dipping a trial GMBO seed in the solution and observing the growth/dissolution processes of the crystal faces at different temperatures. Er,Yb:GBMO bulk crystal was obtained using a well-faceted spontaneous crystal. A GBMO seed was dipped into the solution and overheated by 2–3 °C above the fluxed melt saturation temperature to slightly dissolve the outer surface and ensure defect-free nucleation. Subsequently, the temperature was lowered to the T_{sat} in 1 h. During crystal growth, supersaturation was maintained by cooling down from 0.5 to 2 °C per day. At the end, the crystal was extracted and cooled down to room temperature for several days to prevent cracking due to the thermal shock. All experiments were carried out in a vertical tube furnace, equipped with a CrNi alloy resistive heater and a Proterm-100 precision temperature controller connected with a set of Pt-Rh/Pt thermocouples. The temperature in the furnace working zone was kept with a stability of ±0.1°C.

PXRD studies were performed on a Rigaku MiniFlex-600 powder diffractometer (Rigaku Corp., Tokyo, Japan). PXRD data sets were collected in continuous mode at room temperature (CuK_α radiation) in the range of 2θ = 3–70°, and a scan speed of 5° per minute. The PXRD data were analyzed using the model-biased Le Bail fitting with Yana2006 software [25]. Phases were identified using the Match! software package, version 3.8.1.143 [26], the Crystallographic Open Database (COD) and the ICSD inorganic crystal database [27].

The composition and homogeneity of the Er,Yb:GdMB crystal were studied using the analytical scanning electron microscope (SEM) Leo 1420 VP equipped with the energy dispersive X-ray spectrometer (EDXS) INCA 350. Qualitative analysis was performed on the as-grown face. The sample was fixed on a conductive carbon adhesive tape and covered with a thin layer of carbon to prevent charge accumulation on the crystal during interactions with an electron beam. The Er^{3+} and Yb^{3+} distribution coefficient (K_d) was defined as $K_d = C_{cryst}/C_s$, where C_{cryst} is the Er^{3+} and Yb^{3+} content in the crystal and C_s is the nominal concentration of rare-earth oxides in the initial load.

DSC analysis was performed on a STA 449 F5 Jupiter® (Netzsch, Selb, Germany) to investigate the thermal behavior of the Er,Yb:GMBO crystal. DSC measurements were performed in the temperature range of 50–1200 °C at a heating rate of 20 °C/min under argon gas flow. PtRh20 crucibles with a volume of 85 µL were used for the experiments. The melting point was determined based on the onset temperature of the melting process, with an estimated uncertainty of ±5 °C. To further investigate the thermal behavior, crystals obtained were heat-treated in the range of 1050 °C. The products after annealing were analyzed using the SEM/EDXS method to identify any changes in the phase composition.

Spontaneous Er:Yb:$GdMgB_5O_{10}$ crystals were ground into powder in a corundum mortar to record attenuated total reflection (ATR) spectra. The measurements were carried out on a Fourier spectrometer Bruker IFS 125HR at room temperature in the spectral range 50–2000 cm^{-1} with a resolution of 2 cm^{-1}. In this case, a Mylar beamsplitter and a DTGS receiver were used in the far-infrared (IR) range (50–700 cm^{-1}). A KBr beamsplitter and a DLaTGS receiver were used in the mid-IR range (400–2000 cm^{-1}). In both cases, the source was a Globar.

Transmission spectra in the ultraviolet (UV) range were recorded on an Ocean Insight OCEAN-HDX-UV-VIS spectrometer at room temperature in the spectral range 200–800 nm with a resolution of 0.73 nm. A deuterium lamp was used as a source, and a CCD detector served as a receiver. The measurements used a fragment of a small crystal.

Er,Yb:GMBO is a monoclinic crystal, it is optically biaxial and its optical properties are described along the three main optical axes of the N_g, N_m, and N_p indicatrices [28]. For spectroscopic investigations in polarized light, plates oriented along N_g, N_m, and N_p indicatrices were cut from the Er,Yb:GMBO crystal. The polarized absorption spectra were measured at room temperature by using spectrophotometer Agilent Cary 5000. The absorption cross-section spectra were calculated according to (1):

$$\sigma_{abs}(\lambda) = \frac{k_{abs}(\lambda)}{N} \quad (1)$$

where $k_{abs}(\lambda)$ is the absorption coefficient, N is the concentration of ytterbium for near-1 µm spectra or erbium for near-1.5 µm spectra.

For lifetime measurements, a parametric oscillator based on β-Ba$_2$B$_2$O$_4$ crystal pumped by a 355 nm Nd:YAG laser (third harmonic) was used as an excitation source. Luminescence of the sample was collected from its surface irradiated with an excitation beam, passed through an MDR-12 monochromator, and recorded by an InGaAs photodiode with a preamplifier coupled to a 500 MHz digital oscilloscope.

The energy transfer efficiency was calculated according to (2) [29]:

$$\eta = \tau(1/\tau - 1/\tau_0) \quad (2)$$

where τ is the ytterbium $^2F_{5/2}$-level lifetime in Er,Yb-codoped crystal, and τ_0 is the ytterbium $^2F_{5/2}$-level lifetime in Yb single-doped crystal. For Yb^{3+} lifetime measurements, luminescence kinetics were registered by using immersed Er,Yb:GMBO fine powder that enables minimizing reabsorption influence [30].

The calculation of the stimulated emission spectra in the range near 1.5 µm (transition $^4I_{13/2} \rightarrow ^4I_{15/2}$) was performed by using the integral reciprocity method according to (3) [31] and radiative lifetime of the $^4I_{13/2}$ energy level obtained from the Judd–Ofelt theory in [17].

$$\sigma_{em}^\alpha(\lambda) = \frac{3\exp(-hc/(kT\lambda))}{8\pi n^2 \tau_{rad} c \sum_\gamma \int \lambda^{-4} \sigma_{abs}^\gamma(\lambda) \exp(-hc/(kT\lambda)) d\lambda} \sigma_{abs}^\alpha(\lambda) \quad (3)$$

where $\sigma_{em}(\lambda)$ is the stimulated emission cross-section; α, γ denote the polarization of light; h is Planck's constant; c is the speed of light in a vacuum; k is the Boltzmann constant; T is the temperature of the environment; n is the refractive index of the crystal; τ_{rad} is the radiation lifetime of the $^4I_{13/2}$ energy level of erbium ions; and $\sigma_{abs}(\lambda)$ is the absorption cross-section.

The gain cross-section spectra g(λ) were calculated for different inversion parameters β by using the following Equation (3):

$$g(\lambda) = \beta\, \sigma_{em}(\lambda) - (1-\beta)\sigma_{abs}(\lambda) \quad (4)$$

where β = N_{ex}/N_{tot} is the ratio of the population of excited Er^{3+} ion manifolds to the total erbium ion concentration.

The experimental setup provided in Figure 14 was used to study the laser properties in a continuous-wave mode of operation. A plane–plane N_m-cut Er,Yb:GMBO crystal with a length of 1.5 mm was used as a gain medium. It was coated with anti-reflection coatings at the pump and laser wavelengths and mounted on an aluminum heat sink kept at 20 °C. A continuous-wave fiber-coupled laser diode (Ø 105 µm, NA = 0.22) emitting at a wavelength of 976 nm was used for the longitudinal pumping of the gain medium. The plano–plano cavity with a geometrical cavity length of 4 mm was applied. A single-lens focusing system focused the pump beam into a 120 µm spot inside the laser crystal. Three output couplers with different transmission coefficients at laser wavelengths were used in the laser experiments.

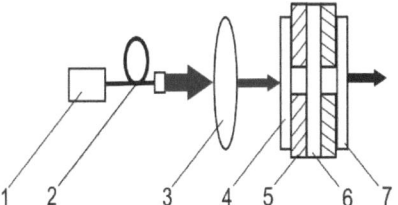

Figure 14. The experimental setup of a continuous-wave diode-pumped Er,Yb:GMBO laser: 1—laser diode; 2—fiber; 3—focusing system; 4—input mirror; 5—copper heat sink; 6—active element; and 7—output coupler.

4. Conclusions

Er^{3+},Yb^{3+}:$GdMgB_5O_{10}$ single crystals with dimensions up to 24 × 15 × 12 mm^3 were successfully grown using an HT-SGDS route from a $K_2Mo_3O_{10}$-based flux system. Detailed study of the growth technique and characterization of Er,Yb:$GdMgB_5O_{10}$ single crystals were carried out. The spectroscopic properties of as-grown crystals were presented. It was shown that Er,Yb:GMBO crystals are characterized by spectroscopic properties to achieve laser operations. A CW diode-pumped Er,Yb:GMBO laser with an output power of about 150 mW and a slope efficiency of 11% was realized at 1568 nm.

Author Contributions: Software, K.N.G., E.V.K. and E.A.V.; validation, K.N.G., V.E.K., V.V.M., E.A.V., E.V.K., D.D.M., N.N.K., E.I.M. and V.L.K.; investigation, K.N.G., V.E.K., V.V.M., E.A.V., E.V.K., D.D.M., N.N.K., E.I.M. and V.L.K.; resources, K.N.G., V.E.K., E.A.V., E.V.K., V.V.M. and N.N.K.; writing—original draft preparation, K.N.G. and E.A.V.; writing—review and editing, K.N.G. and E.A.V.; visualization, E.A.V. All authors have read and agreed to the published version of the manuscript.

Funding: The work was carried out within the framework of the state budget theme AAAA-A16-116033010121-7 of M.V. Lomonosov MSU. ATR and transmission spectra were obtained with the support of the Ministry of Science and Higher Education of the Russian Federation under the program FFUU-2024-004. This research received no external funding.

Data Availability Statement: The original contributions presented in the study are included in the article, further inquiries can be directed to the corresponding author/s.

Conflicts of Interest: The authors declare no conflicts of interest.

References

1. Mutailipu, M.; Poeppelmeier, K.R.; Pan, S. Borates: A Rich Source for Optical Materials. *Chem. Rev.* **2020**, *121*, 1130–1202. [CrossRef]
2. Leonyuk, N.I.; Maltsev, V.V.; Volkova, E.A. Crystal Chemistry of High-Temperature Borates. *Molecules* **2020**, *25*, 2450. [CrossRef] [PubMed]
3. Hinojosa, S.; Meneses-Nava, M.A.; Barbosa-Garcia, O.; Diaz-Torres, L.A.; Santoyo, M.A.; Mosino, J.F. Energy back transfer, migration and energy transfer (Yb-to-Er and Er-to-Yb) processes in Yb,Er:YAG. *J. Lumin.* **2003**, *102–103*, 694–698. [CrossRef]
4. Karlsson, G.; Laurell, F.; Tellefsen, J.; Denker, B.; Galagan, B.; Osiko, V.; Sverchkov, S. Development and characterization of Yb-Er laser glass for high average power laser diode pumping. *Appl. Phys. B* **2002**, *75*, 41–46. [CrossRef]
5. Taccheo, S.; Sorbello, G.; Laporta, P.; Karlsson, G.; Laurell, F. 230-mW diode-pumped single-frequency Er,Yb laser at 1.5 µm. *IEEE Phot. Techn. Let.* **2001**, *13*, 19–21. [CrossRef]
6. Zagumennyi, A.I.; Lutts, G.B.; Popov, P.A.; Sirota, N.N.; Shcherbakov, I.A. The Thermal conductivity of YAG and YSAG laser crystals. *Laser Phys.* **1993**, *3*, 1064–1065.
7. Krankel, C.; Uvarova, A.; Guguschev, C.; Kalusniak, S.; Hülshoff, L.; Tanaka, H.; Klimm, D. Rare-earth doped mixed sesquioxides for ultrafast lasers. *Opt. Mat. Exp.* **2022**, *12*, 1074–1091. [CrossRef]
8. Schweizer, T.; Jensen, T.; Heumann, E.; Huber, G. Spectroscopic properties and diode-pumped 1.6 µm laser performance in Yb-codoped Er:$Y_3Al_5O_{12}$ and Er:Y_2SiO_5. *Opt. Commun.* **1995**, *118*, 557–561. [CrossRef]

9. Bjurshagen, S.; Brynolfsson, P.; Pasiskevicius, V.; Parreu, I.; Pujol, M.C.; Peña, A.; Aguiló, M.; Díaz, F. Crystal growth, spectroscopic characterization, and eye-safe laser operation of erbium- and ytterbium-codopedKLu(WO$_4$)$_2$. *Appl. Opt.* **2008**, *47*, 656–665. [CrossRef]
10. Chen, Y.; Lin, Y.; Huang, J.; Gong, X.; Luo, Z.; Huang, Y. Spectroscopic and laser properties of Er^{3+},Yb^{3+}:LuAl$_3$(BO$_3$)$_4$ crystal at 1.5-1.6 μm. *Opt. Express* **2010**, *18*, 13700–13707. [CrossRef]
11. Huang, J.; Chen, Y.; Gong, X.; Lin, Y.; Luo, Z.; Huang, Y. Spectral and laser properties of Er,Yb:Sr$_3$Lu$_2$(BO$_3$)$_4$ crystal at 1.5–1.6 μm. *Opt. Express* **2013**, *3*, 1885–1892. [CrossRef]
12. Gorbachenya, K.N.; Kisel, V.E.; Yasukevich, A.S.; Deineka, R.V.; Lipinskas, T.; Galinis, A.; Miksys, D.; Maltsev, V.V.; Leonyuk, N.I.; Kuleshov, N.V. Monolithic 1.5 μm Er,Yb:GdAl$_3$(BO$_3$)$_4$ eye-safe laser. *Opt. Mat.* **2019**, *88*, 60–66. [CrossRef]
13. Cheng, W.; Zhang, T.; Jiang, Z.; Huang, G.; Huang, Y.; Li, B.; Lin, Z.; Chen, Y.; Huang, Y.; Lin, Y.; et al. 4.55 W continue-wave dual-end pumping of Er:Yb:YAl$_3$(BO$_3$)$_4$ microchip laser at 1.5 μm. *Appl. Phys. Lett.* **2023**, *123*, 171101. [CrossRef]
14. Chen, Y.; Lin, Y.; Huang, J.; Gong, X.; Luo, Z.; Huang, Y. Fabrication and diode-pumped 1.55 μm continuous-wave laser performance of a diffusion-bonded Er:Yb:YAl$_3$(BO3)$_4$/YAl$_3$(BO$_3$)$_4$ composite crystal. *Opt. Express* **2017**, *25*, 17128–17133. [CrossRef]
15. Gorbachenya, K.N.; Kisel, V.E.; Deineka, R.V.; Yasukevich, A.S.; Kuleshov, N.V.; Maltsev, V.V.; Mitina, D.D.; Volkova, E.A.; Leonyuk, N.I. Continuous-wave laser on Er,Yb-codopedpentaborate crystal. *Devices Methods Meas.* **2019**, *10*, 301–307. [CrossRef]
16. Huang, Y.; Yuan, F.; Sun, S.; Lin, Z.; Zhang, L. Thermal, spectral and laser properties of Er^{3+},Yb^{3+}:GdMgB$_5$O10: A new crystal for 1.5 μm lasers. *Materials* **2018**, *11*, 25. [CrossRef]
17. Huang, Y.; Sun, S.; Yuan, F.; Zhang, L.; Lin, Z. Spectroscopic properties and continuous-wave laser operation of Er^{3+},Yb^{3+}:LaMgB$_5$O$_{10}$ crystal. *J. Alloy Compd.* **2017**, *695*, 215–220. [CrossRef]
18. Chen, Y.; Hou, Q.; Huang, Y.; Lin, Y.; Huang, J.; Gong, X.; Luo, Z.; Lin, Z.; Huang, Y. Efficient continuous-wave diode-pumped Er^{3+}:Yb^{3+}:LaMgB$_5$O$_{10}$ laser with sapphire cooling at 1.57 μm. *Opt. Express* **2017**, *25*, 19320–19325. [CrossRef]
19. Chen, Y.; Huang, Y.; Lin, Z.; Huang, Y. Passively Q-switched Er,Yb:GdMgB$_5$O$_{10}$ pulse laser at 1567 nm. *OSA Contin.* **2019**, *2*, 3598–3603. [CrossRef]
20. Huang, Y.; Lou, F.; Sun, S.; Yuan, F.; Zhang, L.; Lin, Z.; You, Z. Spectroscopy and laser performance of Yb^{3+}:GdMgB$_5$O$_{10}$ crystal. *J. Lumines* **2017**, *188*, 7–11. [CrossRef]
21. Gorbachenya, K.N.; Yasukevich, A.S.; Lazarchuk, A.I.; Kisel, V.E.; Kuleshov, N.V.; Volkova, E.A.; Maltsev, V.V.; Koporulina, E.V.; Yapaskurt, V.O.; Kuzmin, N.N.; et al. Growth and Spectroscopy of Yb:YMgB$_5$O$_{10}$ Crystal. *Crystals* **2022**, *12*, 986. [CrossRef]
22. Sun, S.; Wei, Q.; Li, B.; Shi, X.; Yuan, F.; Lou, F.; Zhang, L.; Lin, Z.; Zhong, D.; Huang, Y.; et al. The YMgB$_5$O$_{10}$ crystal preparation and attractive multi-wavelength emission characteristics of doping Nd^{3+} ions. *J. Mater. Chem. C* **2021**, *9*, 1945–1957. [CrossRef]
23. Wang l Liu, S.; Gou, H.; Chen, Y.; Yue, G.H.; Peng, D.L.; Hihara, T.; Sumiyama, K. Preparation and characterization of the ZnO:Al/Fe$_{65}$Co$_{35}$/ZnO:Al multifunctional films. *Appl. Phys. A Mater. Sci. Process.* **2011**, *106*, 717–723. [CrossRef]
24. Maltsev, V.V.; Mitina, D.D.; Belokoneva, E.L.; Volkova, E.A.; Koporulina, E.V.; Jiliaeva, A.I. Synthesis and flux-growth of rare-earth magnesium pentaborate crystals RMgB$_5$O$_{10}$ (R = Y, Gd, La, Tm and Yb). *J. Cryst. Growth* **2022**, *587*, 126628. [CrossRef]
25. Petricek, V.; Dusek, M.; Palatinus, L. Crystallographic computing system JANA2006: General features. *Z. Krist.* **2014**, *229*, 345–352. [CrossRef]
26. Putz, H.; Brandenburg, K. *Match!—Phase Analysis Using Powder Diffraction, Crystal Impact*; GbR: Bonn, Germany. Available online: https://www.crystalimpact.de/match (accessed on 16 July 2024).
27. *Inorganic Crystal Structure Data Base—ICSD*; Fachinformations Zentrum (FIZ) Karlsruhe: Karlsruhe, Germany, 2021. Available online: https://www.crystallography.net/cod/ (accessed on 16 July 2024).
28. Zhang, J.; Tao, X.; Cai, G.; Jin, Z. Phase relation, structure, and properties of borate MgYB$_5$O$_{10}$ in MgO–Y2O$_3$–B$_2$O$_3$ system. *Powder Diffr.* **2017**, *32*, 97–106. [CrossRef]
29. Burns, P.; Dawes, J.; Dekker, P.; Pipper, J.; Jiang, H.; Wang, J. Optimization of Er,Yb:YCOB for cw laser operation. *IEEE J. Quantum Electron.* **2004**, *40*, 1575–1582. [CrossRef]
30. Sumida, D.S.; Fan, T.Y. Effect of radiation trapping on fluorescence lifetime and emission cross section measurements in solid-state laser media. *Opt. Lett.* **1994**, *19*, 1343–1345. [CrossRef]
31. Yasyukevich, A.S.; Shcherbitskii, V.G.; Kisel, V.E.; Mandrik, A.V.; Kuleshov, N.V. Integral method of reciprocity in the spectroscopy of laser crystals with impurity centers. *J. Appl. Spectr.* **2004**, *71*, 202–208. [CrossRef]

Disclaimer/Publisher's Note: The statements, opinions and data contained in all publications are solely those of the individual author(s) and contributor(s) and not of MDPI and/or the editor(s). MDPI and/or the editor(s) disclaim responsibility for any injury to people or property resulting from any ideas, methods, instructions or products referred to in the content.

MDPI AG
Grosspeteranlage 5
4052 Basel
Switzerland
Tel.: +41 61 683 77 34

Inorganics Editorial Office
E-mail: inorganics@mdpi.com
www.mdpi.com/journal/inorganics

Disclaimer/Publisher's Note: The title and front matter of this reprint are at the discretion of the Guest Editor. The publisher is not responsible for their content or any associated concerns. The statements, opinions and data contained in all individual articles are solely those of the individual Editor and contributors and not of MDPI. MDPI disclaims responsibility for any injury to people or property resulting from any ideas, methods, instructions or products referred to in the content.

www.ingramcontent.com/pod-product-compliance
Lightning Source LLC
LaVergne TN
LVHW070001100526
838202LV00019B/2600